Sir J. William Dawson

The Canadian Ice Age

Being Notes on the Pleistocene Geology of Canada

Sir J. William Dawson

The Canadian Ice Age
Being Notes on the Pleistocene Geology of Canada

ISBN/EAN: 9783337190217

Printed in Europe, USA, Canada, Australia, Japan

Cover: Foto ©berggeist007 / pixelio.de

More available books at **www.hansebooks.com**

Modern Boulder-formation or "Moraine" produced by sea-borne ice, Little Metis, Lower St. Lawrence.

[*From a photograph by Henderson.*]

THE
CANADIAN ICE AGE

BEING NOTES ON THE PLEISTOCENE GEOLOGY OF
CANADA, WITH ESPECIAL REFERENCE TO
THE LIFE OF THE PERIOD AND ITS
CLIMATAL CONDITIONS.

BY

SIR J. WILLIAM DAWSON, C.M.G.

LL.D., F.R.S., F.G.S., &c.

MONTREAL
WILLIAM V. DAWSON
1893

"WITNESS" PRINTING HOUSE,
MONTREAL.

PREFACE.

The facts stated and questions discussed in this work have occupied some portion of the attention of the author since 1855, and he has from time to time published his results in the *Canadian Naturalist and Geologist*, and elsewhere. In 1872 the subject up to that time was summed up in a pamphlet of about 100 pages, entitled "Notes on the Post-pliocene of Canada," now out of print; but since that date no general work on the subject has appeared, though many separate memoirs and papers have been issued by the author and other Canadian geologists.

These papers include a large mass of information bearing on the history of the northern half of the continent of North America in that Ice-age which was in some sense peculiarly its own; and as this material is difficult of access both to geologists and to the general public, no excuse seems necessary for attempting to collect it in a convenient form.

The author has studied the widespread and complex glacial formations of Canada too long to be content to explain them all by one dominant cause, in the rough and ready method employed by so many of his juniors. He has long been convinced that we must take into account the agency both of land ice and sea-borne ice in many forms, along with repeated and complex elevations and depressions of large portions of the continent, in order to account for the effects observed. He is disposed, however, to seek for the causes of changes of climate rather in geological and geographical agencies than in astronomical vicissitudes, some of which are too slow and uncertain in their operation, and others altogether conjectural. Such views are less sensational than those which invoke vast and portentous exaggerations of individual phenomena, but they are likely, in the end, to commend themselves to serious thinkers, especially when they are confirmed by the facts observed in the regions which are, of all others, best suited for the study both of extinct and recent ice-action.

The basis of the work rests on the observations of the author, but reference will be made to the large and important contributions of Dr. G. M. Dawson, Dr. Bell, Dr. Ells, Mr. Whiteaves, Mr. Chalmers, Mr. Low, Dr. Spencer, Mr. McConnell, Mr. Richardson, Prof. Yule Hind, Lieut.-Col. Grant, Dr. G. J. Hinde, and others, whose names will be found in the subsequent pages as workers in the Pleistocene geology of Canada.

J. WM. DAWSON.

McGILL COLLEGE,
          MONTREAL, 1893.

CONTENTS.

LIST OF ILLUSTRATIONS.

Glaciated Laurentian hills, north shore of St. Lawrence, opposite Kamouraska.

THE

ICE AGE IN CANADA.

CHAPTER I.

HISTORICAL NOTICES.

Canada presents unsurpassed facilities for the study of the pleistocene deposits. Extending across the American Continent from the Atlantic to the Pacific, in the widest part of that continent, and reaching from the latitude of 45° to the polar regions, possessing great plains covered with drift material, and mountainous districts heavily marked with the action of land ice, and having in many places abundance of fossil remains in its more recent deposits, it has the same relative facilities for the study of this later geological period that it has for the earlier Laurentian; and it has been one of the objects of the ambition of the writer for the last thirty years, to do a little toward making it a typical region for the Pleistocene, as Logan has for the Laurentian. I shall endeavour, therefore, to sketch the Pleistocene as it appears in Canada.

In making this attempt, I have all along felt compelled

2

by facts to insist on the following great leading principles
as to the glacial period and its causes :—

1. The phenomena are not to be explained by any one
cause or by any one great all-embracing hypothesis, but
by a more active and extensive operation of many of the
ordinary causes still existing in the more northern regions.

2. The astronomical changes which have been invoked
to account for cold climate, not excepting those advocated
by Croll and Ball, are incapable of fully explaining the
facts as now actually ascertained.

3. There has not been at any time a polar ice cap, and
the theory of great continental ice sheets covering the
northern parts of the two great continents is also baseless.

4. The phenomena indicate the action of local mountain
glaciers of great volume, along with that of floating ice
in various forms, and this more especially in periods of
subsidence of the land.

5. The cold climate of the glacial period was mainly a
result of peculiar geographical conditions and a different
distribution of ocean currents, and was not so much
characterised by general low temperature as by the local
occurrence of extreme evaporation and condensation.

6. The close of the glacial period is not very remote,
and cannot have antedated by many centuries or mil-
lenniums the first appearance of man, as known to us in
history.

These theses I have maintained in papers whose dates
reach back to 1855, and in addresses delivered to the
British and American Associations and to the Natural
History Society of Montreal, and in popular works on
geology. They still appear to me to be true, notwithstanding the wave of extreme glacial ideas that has been
passing over the world. But I am glad to see that a

reaction to better views has begun to set in, and that geologists are more disposed than formerly to restrict their speculations within the limits of physical possibility. We shall see evidence of this in the sequel; but before referring to the conclusions of others, I may be pardoned for giving a sketch of the progress of opinion, as it has presented itself in connection with my own work.

When I first entered on the study of these deposits in Nova Scotia, in the year 1841 and subsequent years, my guide and instructor was the great apostle of moderate uniformitarianism, that is, of rational geology, Sir Charles Lyell. His views as to the combined agency of land ice or glaciers, of floating fragments of glaciers or ice-bergs, and of field ice, are, or ought to be, well known; but I must say that they have often been unfairly stated. Lyell well knew the nature and work of glaciers in so far as ascertained in his time. He had also collected a large amount of information as to the conveyance of boulders, etc., by ice-bergs, and the formation of submarine glacial deposits thereby. Lastly, he had profited by the observations of the Arctic voyagers, and by those of Bayfield in the river Saint Lawrence, so as to appreciate the great carrying and erosive power of heavy field ice. His general theory of the glacial age was based on all these factors, along with the gradual depression and re-elevation of the continents in the pleistocene period. I confess that I still adhere to his views in these respects, with only such modification as to the relative value of particular and local causes, as the observations and reading of fifty years have necessitated.

My own conclusions with reference to the phenomena observed in the Maritime Provinces of Canada, were expressed for the first time in the first edition of "Acadian

Geology" in 1854–5. The following extract will show that
they were formed very closely on the Lyellian doctrine
of "modern causes":—

"If we ask what has been the origin of this great mass
of shifted and drifted material which overspreads the
surface, not only of the province we are now describing,
but the greater part of the land of the northern hemisphere, we raise one of the most vexed questions of modern
geology. In reasoning, however, on this subject as regards
Nova Scotia, I have the advantage of appealing to causes
now in operation within the country. In the first place,
it may at once be admitted that no such operations as
those which formed the drift are now in progress on the
surface of the land, so that the drift is a relic of a past state
of things, in so far at least as regards the localities in which
it now rests. In the next place, we find, on examining
the drift, that it strongly resembles, though on a greater
scale, the effects now produced by frost and floating ice.
Frost breaks up the surface of the most solid rocks, and
throws down cliffs and precipices. Floating ice annually
takes up and removes immense quantities of loose stones
from the shores, and deposits them in the bottom of the
sea or on distant parts of the coasts. Very heavy masses
are removed in this way. I have seen in the strait of
Canseau large stones ten feet in diameter, that had been
taken from below low water mark and pushed up upon
the beach. Stones so large that they had to be removed
by blasting, have been taken from the base of the cliffs at
the Joggins and deposited off the coal-loading pier, and I
have seen resting on the mud flats at the mouth of the
Petitcodiac river a boulder at least eight feet in length,
that had been floated by the ice down the river. Another
testimony to the same fact is furnished by the rapidity

with which huge piles of fallen rock are removed by the floating ice from the base of the trap cliffs of the bay of Fundy. Let us suppose, then, the surface of our province, while its projecting rocks were still uncovered by surface deposits, exposed for many successive centuries to the action of alternate frosts and thaws, the whole of the untravelled drift might have been accumulated on its surface. Let it then be submerged until its hill-tops should become islands or reefs of rock in a sea loaded in winter and spring with drift ice, floated along by currents, which, like the present arctic current, would set from N.E. to S.W. with various modifications produced by local causes. We have in these causes ample means for accounting for the whole of the appearances, including the travelled blocks and the scratched and polished rock-surfaces."

This was written, it may be observed, thirty-five years ago, and with reference to the phenomena presented by southern New Brunswick and Nova Scotia, where there is little if any evidence of glacier action.

When, in the autumn of 1855, my residence was transferred to Montreal, my attention was necessarily devoted to the pleistocene deposits of Central Canada, and I asked Sir W. E. Logan, then Director of the Geological Survey, to place in my hands, as an amateur, the pleistocene geology of this field, which he readily consented to do, as no one connected with the survey was specially cultivating it at the time. I proceeded, in the first instance, to explore the stratigraphical arrangement and fossils of the deposits, dividing the former into the three groups of Boulder Clay, Leda Clay and Saxicava Sand, and raising the known species of fossils in a few years from a very small number to about 200. Notices of these researches were published from time to time in the "Canadian Naturalist and Geologist." When, in 1863, Sir William issued his "Geology of Canada," I was much occupied with college work, and felt that the subject was too immature to admit of full treatment, but placed in his hands my notes up to that date to aid in his chapter on "Superficial Geology," in which they were incorporated, though in an imperfect manner. Subsequently, in 1872, I collected all my papers up to that date in a little volume entitled "Notes on the Post-pliocene Geology of Canada," now out of print, though most of its material is to be found in the earlier volumes of the "Canadian Naturalist and Geologist." This work I have made the basis of subsequent publications, adding new material as it occurred, and publishing the whole in the same periodical and its continuation, the "Canadian Record of Science." The present work is a new and enlarged edition of these "Notes" of 1872.

Since my work in this field began, the subject has assumed many new phases. The important studies of the Swiss glaciers, by Forbes, Agassiz and others, attracted the attention of geologists almost to the exclusion of other factors. The bold, I may venture to say extreme, views of my friends, Ramsay and Geikie, have given a tone to the work of English geologists, while a like influence has been exercised in America by Agassiz and Dana. Thus, in later years, what I must regard as extravagant theories of land glaciation have gained an educational and official currency both in England and America. Only recently the pendulum has begun to swing in the other direction, and the extreme theories of glacier action to relax their hold. The time is, therefore, perhaps a favourable one to advocate moderate and rational views, and perhaps to prevent an undue reaction in the direction opposite to that lately prevalent.

I trust I shall not be accused of egotism if I present these moderate views, in the first instance, in the form of extracts from publications dating some of them nearly thirty years ago.

In a paper published in the " Canadian Naturalist" in 1860, and specially devoted to the description of glacial phenomena in Labrador, Maine, etc., the following words occur with reference more particularly to the climate of the Pleistocene, and are here given without alteration.*

"Everyone knows that the means and extremes of annual temperature differ much on the opposite sides of the Atlantic. The isothermal line of 40°, for example, passes from the south side of the gulf of St. Lawrence, skirts Iceland and reaches Europe near Drontheim in Norway. This fact, apparent as the result of observations on the temperature of the land, is equally evidenced by the inhabitants and physical phenomena of the sea. A large proportion of the shell-fish inhabiting the gulf of St. Lawrence and the coast thence to Cape Cod, occur on both sides of the Atlantic, but not in the same latitudes. The marine fauna of Cape Cod is parallel in its prevalence of boreal forms with that of the south of Norway. In like manner the descent of icebergs from the north, the freezing of bays and estuaries, the drifting and pushing of stones and boulders by ice, are witnessed on the American coast in a manner not paralleled in corresponding latitudes in Europe. It follows from this that a collection of shells from any given latitude on the coasts of Europe or

---

* Where anything new is introduced into these extracts, it is placed in brackets, thus,—[. . .]

America, would bear testimony to the existing difference
of climate. The geologist appeals to the same kind of
evidence with reference to the climate of the later
tertiary period, and let us enquire what is its testimony.

"The first and most general answer is that the pleistocene climate was colder than the modern. The proof of
this in western Europe is very strong. The marine
fossils of this period in Britain are more like the existing
fauna of Norway or of Labrador than the present fauna
of Britain. Great evidences exist of driftage of boulders
by ice, and traces of glaciers on the higher hills. In
North America the proofs of a rigorous climate, and
especially of the transport of boulders and other materials
by ice, are equally good, and the marine fauna all over
Canada and New England is of boreal type.

"Admitting, however, that a rigorous climate prevailed
in the pleistocene period, it by no means follows that the
change has been equally great in different localities. On
the contrary, while a great and marked revolution has
occurred in Europe, the evidences of such change are very
much more slight in America. In short, the causes of
the coldness of the pleistocene seas to some extent still
remain in America, while they have disappeared or have
been modified in Europe.

"If we inquire as to these causes as at present existing,
we find them in the distribution of ocean currents, and
especially in the great warm current of the gulf stream,
thrown across from America to Europe, and in the arctic
currents bathing the coasts of America. In connection
with these we have the prevailing westerly winds of the
temperate zone, and the great extent of land and shallow
seas in northern America. Some of these causes are
absolutely constant. Of this kind is the distribution of

the winds, depending on the earth's temperature and rotation. The courses of the currents are also constant, except in so far as modified by coasts and banks; and the direction of the drift-scratches and transport of boulders in the Pleistocene both of Europe and America, show that the arctic currents at least have remained unchanged. But the distribution of land and water is a variable element, since we know that in the period in question nearly all northern Europe, Asia and America were at one time or another under the waters of the sea, and it is consequently to this cause that we must mainly look for the changes which have occurred.

"Such changes of level must, as has been long since shown by Sir Charles Lyell, modify and change climate. Every diminution of the land in arctic America must tend to render its climate less severe. Every diminution of land in the temperate regions must tend to reduce the mean temperature. Every diminution of land anywhere must tend to diminish the extremes of heat and cold; and the condition of the southern hemisphere at present shows that the submergence of the great continental masses would lower the mean temperature, but render the climate much less extreme. Glaciers might then exist in latitudes where now the summer heat would suffice to melt them, as Darwin has shown that in South America glaciers extend to the sea level in latitude 46° 50′; and at the same time the ice would melt more slowly and be drifted farther to the southward. [In the southern hemisphere, indeed, a glacial period of a peculiar kind exists at present, since there is an ice-bound antarctic continent 2,000 miles in diameter and boulder-drift extending from it half-way to the equator.] Any change that tended to divert the arctic currents from our

coasts would raise the temperature of their waters. Any
change that would allow the equatorial current to pursue
its course through to the Pacific or along the great inland
valley of North America, would reduce the British seas to
a boreal condition.

"The boulder formation and its overlying fossiliferous
beds prove, as I have in previous papers endeavoured to
explain with regard to Canada, and as has been shown by
other geologists in the case of other regions, that the land
of the northern hemisphere underwent in the later
tertiary period a great and gradual depression and then
an equally gradual elevation. Every step of this process
would bring its modifications of climate, and when the
depression had attained its maximum there probably was
as little land in the temperate regions of the northern
hemisphere as in the southern now [while that which
remained above water was high and mountainous].* This
would give a low mean temperature and an extension to
the south of glaciers, more especially if at the same time
a considerable arctic continent remained above the waters
[as a gathering ground], as seems to be indicated by the
effects of extreme marine glacial action on the rocks
under the boulder clay. These conditions, actually indicated by the phenomena themselves, appear quite sufficient
to account for the coldness of the seas of the period, and
the wide diffusion of the gulf stream caused by the
subsidence of American land, or its entire diversion into
the Pacific basin,† would give that assimilation of the

---

* The important question of differential elevation has been solved in
great part since this was written, and would much strengthen the
argument.

† This is often excluded from consideration, owing to the fact that
the marine fauna of the gulf of Mexico differs so much from that of the

American and European climates so characteristic of the time. The climate of western Europe, in short, would, under such a state of things, be greatly reduced in mean temperature, the climate of America would suffer a less reduction of its mean temperature, but would be much less extreme than at present; the general effect being the establishment of a more equable but lower temperature throughout the northern hemisphere.

"The establishment of the present distribution of land and water, giving to America its extreme climate, leaving its seas cool and throwing on the coasts of Europe the heated water of the tropics, would thus affect but slightly the marine life of the American coast, but very materially that of Europe, producing the result already referred to, that our Canadian pleistocene fauna differs comparatively little from that now existing in the gulf of St. Lawrence, though in so far as any difference subsists it is in the direction of an arctic character. The changes that have occurred are perhaps all the less that so soon as the Laurentide hills to the north of the St. Lawrence valley emerged from the sea, the coasts to the south of these hills would be effectually protected from the heavy northern ice drifts and from the arctic currents, and would have the benefit of the full action of the summer

Pacific coast; but, on the other hand, the occurrence of many species common to the two sides implies a connection in comparatively recent times, and similar evidence is afforded by the modern deposits of the Isthmus.

Upham reports from Dr. Maack, in the reports of explorations for the Panama canal, the fact that on the watershed between the Atlantic and Pacific a " vast area " of the Isthmus is occupied with " late tertiary " beds holding shells of living species. This would confirm the supposition based on the grounds that a passage across the Isthmus existed in pleistocene times.—American Geologist, December, 1890.

heat, advantages which must have existed to a less extent in western Europe.*

"It is farther to be observed that such subsidence and elevation would necessarily afford great facilities for the migration of arctic marine animals, and that the difference between modern and pleistocene faunas must be greatest in those localities to which the animals of temperate regions could most readily migrate after the change of temperature had occurred."

In an address delivered in 1864 † as retiring president of the Natural History Society of Montreal, the relative importance of land-ice and sea-borne ice is referred to in the following terms, in connection with the then recent appearance of Logan's "General Report on the Geology of Canada," published in 1863 :—

"There is another subject of great geological importance on which the publication of this report enables strong ground to be taken. I refer to the conditions under which the *boulder-drift* of Canada was deposited. It has been customary to refer this to the action of ice-laden seas and currents, on a continent first subsiding and then re-elevated. But this opinion has recently been giving way before a re-assertion of the doctrine that land-glaciers have been the principal agents in the distribution of the boulder-drift, and in the erosions with which it was accompanied. I confess that I have steadily rejected this last doctrine; being convinced that insuperable physical

---

* One cannot be too emphatic in insisting on the fact that, in North America, throughout geological time, movements of subsidence which threw open the interior plains to the arctic currents produced refrigeration, while those that produced a great mediterranean sea, open to the south and closed on the north, introduced mild climates.

† Canadian Naturalist, 1864.

and meteorological objections might be urged against it, and that it was not in accordance with the facts which I had myself observed in Nova Scotia and in Canada. The additional facts contained in the present report enable me to assert with confidence, though with all humility, that glaciers could scarcely have been the agents in the striation of Canadian rocks, the transport of Canadian boulders, or the excavation of Canadian lake-basins [except in the great mountain ranges of the continent].

"The facts to be accounted for are the striation and polishing of rock-surfaces, the deposit of a sheet of unstratified clay and stones, the transport of boulders from distant sites lying to the northward, and the deposit on the boulder-clay of beds of stratified clay and sand, containing marine shells. The rival theories in discussion are—*first*, that which supposes a gradual subsidence and re-elevation, with the action of the sea and its currents, bearing ice at certain seasons of the year; and, *secondly*, that which supposes the North American plateau to have been covered with a sheet of glacier several thousands of feet thick.

"The last of these theories, without attempting to undervalue its application to such regions as those of the Alps or of Spitzbergen or Greenland, has appeared to me inapplicable to the drift-deposits of eastern America, for the following among other reasons:

"1. It requires a series of suppositions unlikely in themselves and not warranted by facts. The most important of these is the coincidence of a wide-spread continent and a universal covering of ice in a temperate latitude. In the existing state of the world, it is well known that the ordinary conditions required by glaciers in temperate latitudes are elevated chains and peaks extending above the snow-line; and that cases in which, in such latitudes, glaciers extend nearly to the sea-level, occur only where the mean temperature is reduced by cold ocean currents approaching to high land, as for instance in Terra del Fuego and the southern extremity of South America. But the temperate regions of North America could not be covered with a permanent mantle of ice under the existing conditions of solar radiation; for even if the whole were elevated into a table-land, its breadth would secure a sufficient summer heat to melt away the ice, except from high mountain peaks.

" 2. It seems physically impossible that a sheet of ice, such as that supposed, could move over an uneven surface, striating it in directions uniform over vast areas, and often different from the present inclinations of the surface. Glacier-ice may move on very slight slopes, but it must follow these [since gravitation, along with the more or less plastic nature of the ice, has been shown to be the cause of its motion]; and the only result of the immense accumulation of ice supposed, would be to prevent motion altogether by the want of slope or the counteraction of opposing slopes, or to induce a slight and irregular motion toward the margins or outward from the more prominent protuberances.

" It is to be observed, also, that, as Hopkins has shown, it is only the *sliding* motion of glaciers that can polish or erode surfaces, and that any internal changes resulting from the mere weight of a thick mass of ice resting on a level surface, could have little or no influence in this way.

" 3. The transport of boulders to great distances, and the lodgment of them on hill-tops, could not have been occasioned by glaciers. These carry downward the blocks that fall on them from wasting cliffs. But the universal glacier supposed could have no such cliffs from which to collect; and it must have carried boulders for hundreds of miles, and left them on points as high as those they were taken from. On the Montreal Mountain, at a height of 600 feet above the sea, are huge boulders of gneiss from the Laurentide hills, which must have been carried 50 to 100 miles from points of scarcely greater elevation, and over a valley in which the striæ are in a direction nearly at right angles with that of the probable driftage of the boulders. Quite as striking examples occur in many parts of this country.* It is also to be observed that boulders, often of large size, occur scattered through the marine stratified clays and sands containing sea-shells; and whatever views may be entertained as to other boulders, it cannot be denied that these have been borne by floating ice. Nor is it true, as has been often affirmed, that the boulder-clay is destitute of marine fossils. At Murray Bay, Riviere du Loup and St. Nicholas, on the St. Lawrence, and also at Cape Elizabeth, near Portland, there are tough stony clays of the nature of true "till," and in the lower part of the drift, which contain numerous marine shells of the usual pleistocene species.

"4. The pleistocene deposits of Canada, in their fossil remains and general character, indicate a gradual elevation from a state of depression, which, on the evidence of fossils, must have extended to at least 500 feet, and on that of far-travelled boulders, to nearly ten times that amount, while there is nothing but the boulder-clay to represent the previous subsidence, and nothing whatever

* The same fact, and to heights still greater, has been shown by Ells and others to hold of the hills of the Eastern Townships of Canada, and by Chalmers in Eastern Quebec and in New Brunswick.

to represent the supposed previous ice-clad state of the
land, except the scratches on the rock-surfaces, which
must have been caused by the same agency which
deposited the boulder-clay.*

" 5. The peat deposits with fir-roots, found below the
boulder-clay in Cape Breton, the remains of plants and
land-snails in the marine clays of the Ottawa, and the
shells of the St. Lawrence clays and sands, show that the
sea at the period in question had nearly the temperature
of the present arctic currents of our coasts, and that the
land was not covered with ice, but supported a vegetation
similar to that of Labrador and the north shore of the St.
Lawrence at present. This evidence refers not to the
later period of the mammoth and mastodon, when the
re-elevation was perhaps nearly complete, but to the
earlier period contemporaneous with, or immediately following, the supposed glacier-period. In my former
papers on the Pleistocene of the St. Lawrence, I have
shown that the change of climate involved is not greater
than that which may have been due to the subsidence of
land, and to the change of course of the equatorial and
arctic currents, actually proved by the deposits themselves.

" These objections might be pursued to much greater
length ; but enough has been said to show that there are
in the case of north-eastern America, strong reasons
against the existence of any such period of extreme
glaciation as supposed by many geologists; and that if we
can otherwise explain the rock striation and polishing,
and the formation of fiords and lake-basins, the strong

---

* This was intended to apply to the valley of the St. Lawrence, not
to the mountainous regions having local glaciers.

points with these theorists, we can dispense altogether with the portentous changes in physical geography involved in their views, and which are not necessary to explain any of the other phenomena."

The address then proceeds to deal with these points in the manner to be stated in a subsequent chapter.

Two years later, in 1866, I prepared an account of all that was known up to that date of the fossil plants of the mid-pleistocene beds, the Leda clay of the Ottawa and St. Lawrence valleys, equivalent in age to the so-called inter-glacial beds of western Canada. In this paper the following summary is given of the climatal conditions indicated :—

"None of the plants from these mid-glacial beds is properly arctic in its distribution, and the assemblage may be characterized as a selection from the present Canadian flora of some of the more hardy species having the most northern range. Green's Creek is in the central part of Canada, near to the parallel of 46°, and an accidental selection from its present flora, though it might contain the same species found in the nodules, would certainly include with these, or instead of some of them, more southern forms. More especially the balsam poplar, though that tree occurs plentifully on the Ottawa, would not be so predominant. But such an assemblage of drift plants might be furnished by any American stream flowing in the latitude of 50° to 55° north. If a stream flowing to the north, it might deposit these plants in still more northern latitudes, as the McKenzie river does now. If flowing to the south, it might deposit them to the south of 50°. In the case of the Ottawa, the plants could not have been derived from a more southern locality, nor probably from one very far to the north.

3

We may, therefore, safely assume that the refrigeration indicated by these plants would place the region bordering the Ottawa in nearly the same position with that of the south coast of Labrador fronting on the gulf of St. Lawrence at present. The absence of all the more arctic species occurring in Labrador, should perhaps induce us to infer a somewhat more mild climate than this [and also indicates the probable existence of the arctic flora to the northward throughout the pleistocene period]."

" The moderate amount of refrigeration thus required, would, in my opinion, accord very well with the probable conditions of climate deducible from the circumstances in which the fossil plants in question occur. At the same time when they were deposited, the sea flowed up the Ottawa valley to a height of 200 to 400 feet above its present level, and the valley of the St. Lawrence was a wide arm of the sea, open to the arctic current. Under these conditions the immense quantities of drift ice from the northward, and the removal of the great heating surface now presented by the low lands of Canada and New England, must have given for the Ottawa coast of that period a summer temperature very similar to that at present experienced on the Labrador coast, and with this conclusion the marine remains of the Leda clay, as well as the few land mollusks whose shells have been found in the beds containing the plants, and which are species still occurring in Canada, perfectly coincide.

" The climate of that portion of Canada above water at the time when these plants were imbedded, may safely be assumed to have been colder in summer than at present, to an extent equal to about 5° of latitude, and this refrigeration may be assumed to correspond with the requirements of the actual geographical changes implied.

In other words, if Canada was submerged until the Ottawa valley was converted into an estuary inhabited by species of *Leda*, and frequented by capelin, the diminution of the summer heat consequent on such depression, would be precisely suitable to the plants occurring in these deposits, without assuming any other cause of change of climate."

This extract, referring as it does to the evidence of plants, reminds us of the contrast between the Pleistocene and the warm climate of the early Eocene and later Cretaceous, when warm temperate plants could flourish as far north as Greenland. The reason is seen in our comparative maps of the Cretaceous and Pleistocene of Canada. The conditions presented in the latter show the greatest possible facilities for the transference of arctic ice to temperate latitudes, and its accumulation therein, while leaving the extreme arctic comparatively free of ice. Such conditions are the reverse of those in the early Eocene, when the interior of the continent was occupied with a warm mediterranean sea, shielded from the arctic ice. Thus the known geographical condition of the Pleistocene harmonize with rational views as to the causes and extent of the refrigeration.

Lastly, in my address to the Natural History Society of Montreal, in 1873, immediately after the publication of the " Notes," above referred to, the following reference to the position of the question occurs :—

" In the memoir in the Journal of the Natural History Society already referred to, I have re-asserted and supported by many additional proofs the theory of the combined action of floating ice and glaciers in the production of our Canadian boulder clay and other superficial deposits, which I have for many years maintained, in opposition to the views of the extreme glacialists. It is matter of gratification to me to find, in connection with this, that researches in other regions are rapidly tending to overthrow extreme views on the subject, and to restore this department of geological dynamics more nearly to the domain of ordinary existing causes. Whymper, Bonney, and other Alpine explorers, have ably supported in England the conclusion which, after a visit to Switzerland in 1865, I ventured to affirm here, that the erosive power of glaciers is very inconsiderable. Mr. Milne Home, Mr. McIntosh, and others, have combated the prevalent notions of a general glacier in England and Scotland. Mr. James Geikie, a leading advocate of land glaciers, has been compelled to admit that marine beds are interstratified with the true boulder clay of Scotland, and consequently to demand a succession of elevations and depressions in order to give any colour to the theory of a general glacier. The idea of glacier action as means of accounting for the drifts of Central Europe and of Brazil seems to be generally abandoned. Lastly, in a recent number of the "American Journal of Science," Prof. Dana has admitted the necessity, in order to account for land glaciation of the hills of New England [by a continental glacier], of supposing a mountain range or table land of at least 6,000 feet in height, to have existed between the St. Lawrence and Hudson's Bay, while in addition to the imaginary N. W. & S. E. glacier, flowing from this immense and improbable mass, there must have been a transverse glacier running beneath it up the valley of the St. Lawrence. Such demands amount, in my judgment, to a virtual abandonment of the theory of other than large local glaciers in America in the pleistocene period. Thus there are cheering indications that the world-enveloping glacier, which has so long spread its icy pall over the geology of the later tertiary periods, is fast melting away before the sunshine of truth."

Perhaps I was a little too sanguine as to the rapidity of the process, and did not make allowance for that chilling current of popular text-books and official influence which has so much retarded the final melting of the great continental glacier and the polar ice cap.

The following citations, however, from very recent publications show that my forecast of the course of opinion was not altogether wrong, and that we may hope for still better things in the future.

Sir Robert Ball's recent attempt to rehabilitate Croll's ingenious astronomical theory of the glacial age will not assist to restore its waning fortunes, but brings out the fact that this ingenious theory was essentially defective.* He shows that Croll reasoned on a mistaken assumption that the earth receives equal amounts of heat when in perihelion and aphelion passage, whereas the difference is as much as 26 per cent., and consequently at long intervals there might occur periods of great coldness in one hemisphere at a time. The interval of time, however, is too long, even on Ball's theory, and the fact that the ice of the glacial period radiated from points consider-ably south of the polar circle, tells of the dominancy of terrestial conditions. This new astronomical theory will therefore fail to affect geological conclusions, and its weak points have already been pointed out by geological reasoners. It may, in short, be held to have given a death blow to the theory of astronomical causes of the glacial period known to us in geology.

In a recent paper, Mr. Warren Upham, one of the most

---

* "The Cause of an Ice Age," London, 1892.

active and voluminous of American glacialists,* adduces a
long array of citations from Geikie, Woeickkoff, Winchell,
Gilbert, Andrews, Wright, Mackintosh, Logan, Spencer
and others on facts showing the recency of the glacial
age, which he holds could not have been more remote
from our time than from 7,000 to 10,000 years. He
further concludes, from the observations of Dr. G. M.
Dawson and others, that "the cause of the glacial period
was great uplifts of the glaciated areas," accompanied and
followed by great and unequal subsidence; and he brings
evidence to show that the passage of the equatorial
current into the Pacific, in consequence of the subsidence
of Central America, diverted the warm waters of the
Gulf Stream from the Atlantic. Claypole, in a review of
the Astronomical Theory,† adduces a great number of
facts in evidence of the recency of the glacial period and
of its origin from terrestrial causes. In a later paper,‡
in which he endeavours to harmonise the doctrines of
great ice-sheets and local glaciers, he admits that the
evidence from both Europe and North America "opposes
the theory of a great polar ice cap while favouring that of
a number of separate radiants."

The great subsidence of the pleistocene period is
emphasised by recent discoveries of high-level glacial
gravels in England, which are described by Mr. A. C.
Nicholson, and which show a depression in that country
to the level of more than 1,300 feet.§

---

* American Geologist, December, 1890.

† Trans. Ed. Geol. Society, Vol. V.

‡ "Glacial Radiants," American Geologist, 1889.

§ "High Level Gravels at Gloppa," Journal Geological Society,
February, 1892.

It is true that many of these writers still speak of an ice-sheet as possibly extending across North America; but their own statements, as well as the now universally understood fact that the interior plateaux of great continents cannot under any circumstances receive snow enough to breed great glaciers, point to entirely different conclusions.

Lastly, in evidence of the gradual return to more moderate opinions, I may quote the words of a veteran glacialist, and a man of wide knowledge and power of thought, who has recently passed away. Alexander Winchell thus refers to Canadian conclusions in a paper on " Recent Views about Glaciers ":

" Now, the most unexpected results of all the recent researches appear to be these: There has been no continental glacier. There has been no uniform southerly movement of glacier masses. There has been no persistent declivity, as a *sine qua non* down which glacier movements have taken place. The continuity of the supposed continental glacier was interrupted in the regions of the dry and treeless plains of the West; and in the interior and Pacific belts of the continent, within the United States, ancient glaciation was restricted to the elevated slopes. A non-glaciated belt, lying a few miles east of the foothills of the Rockies, extends all the way to the unglaciated arctic region. . . . .

" Another unexpected result of continent-wide observation is the discovery of glacial striations tending in all directions from two general centres. One of these is a north-eastern centre of glaciation, which Dr. G. M. Dawson proposes to call the ' Laurentide ' centre; the other is a north-western centre, which he calls ' Cordilleran.' The Cordilleran glacier lay between the range of the

Rocky Mountains proper on the east, and the coast ranges on the Pacific, and stretched from Victoria to latitude 59°. All this region, consisting of the mountain ranges and of the elevated plateau between them, was once occupied by a confluent glacier 1,200 miles long and 400 miles wide. The main gathering-ground, however, was between the 55th and 59th parallels. Thence the ice flowed northward 350 miles, and southward about 600 miles, diverging, in the intervening distance, easterly and westerly."

The Laurentide glacier had for its centre of distribution the highlands called the Laurentian mountains, one arm of which lies between Hudson's Bay and the great plains of British America. Hitherto, most of our knowledge of glacier movements has pertained to the southward-moving portions of this great sheet, but we now know that the remoter parts moved north and north-west. Dr. Bell has shown that there was also a centre of distribution in the peninsula of Labrador, from which movements radiated east, west, south and north, but without reaching the coast northward. This, however, may not have been an independent centre of snowy accumulation, as one arm of the Laurentian ridge extends through Labrador.

Appended to this chapter is a list of the several papers referred to above and in the following chapters, part of which have appeared in the "Canadian Naturalist and Geologist," and its successor, the "Canadian Record of Science."

References to memoirs by other authors will be found in their proper places in the subsequent chapters.

LIST OF PAPERS ON PLEISTOCENE OF CANADA.

(1.) Notice of the Pleistocene Geology of Nova Scotia.—" Acadian Geology," 1855.
(2.) On the Newer Pliocene and Post-pliocene of the Vicinity of Montreal.—Canadian Naturalist, 1857.
(3.) Additional Notes on the Post-pliocene Deposits of the St. Lawrence Valley.—*Ib.* 1859.
(4.) On the climate of Canada in the Post-Pliocene Period.—*Ib.* 1860.
(5.) On Post-Tertiary Fossils from Labrador.—*Ib.* 1860.
(6.) On the Geology of Murray Bay (Part 3, Post-pliocene Deposits.—*Ib.* 1861.
(7.) Address as President of the Natural History Society of Montreal.—*Ib.* 1864.
(8.) On the Post-pliocene Deposits of Rivière du Loup and Tadoussac.—*Ib.* 1865.
(9.) Comparison of the Icebergs of Belle-isle and the Glaciers of Mont Blanc, with reference to the Boulder-clay of Canada.—*Ib.* 1866.
(10.) On the Evidence of Fossil Plants as to the Post-pliocene climate of Canada.—*Ib.* 1866.
(11.) Report of Geology of Prince Edward Island.—1871.
(12.) Notes on the Post-pliocene Geology of Canada, pp. 112 and Figures.—Montreal, 1872. Also Canadian Naturalist, N.S., Vol. VI.
(13.) Address to Natural History Society of Montreal.—1874.
(14.) Note on a Fossil Seal.—Canadian Naturalist, 1877.
(15.) Supplement to Acadian Geology.—1878.
(16.) Notice of Woeickkoff on Glaciers.—Canadian Naturalist, Vol. X.
(17.) Canadian Pleistocene.—London Geological Magazine, 1883.
(18.) On the Skeleton of a Whale found at Smith's Falls, Ontario.—Canadian Naturalist, 1883.
(19.) Geology of the line of the Canadian Pacific Railway.—Journal Geological Society, London, 1884.
(20.) Boulder Drift and Sea Margins at Little Metis.—Canadian Record of Science, 1886. ·
(21.) Pleistocene Fossils from Anticosti.—*Ib.* 1886.
(22.) *Balanus Hameri* and Varieties of *Mya arenaria* and *M. truncata* in the Pleistocene.—*Ib.* 1889.
(23.) On a Fossil Fish and Marine Worm.—*Ib.* 1890.
(24.) The Pleistocene Flora of Canada (with descriptions of the plants by Prof. Penhallow).—Bul. Geol. Society of America, 1890.

Much of the matter contained in these detached publications now requires revision, more especially the lists of fossils; and many additional facts have accumulated. I purpose, therefore, now to summarize the facts and conclusions of my previous papers and to unite them with the new facts, so as to present as complete a view as possible of the geology of the superficial deposits of Canada. I shall also prepare a complete list of the fossils up to date, with revised nomenclature and synonymy. In this last part of the work I have been aided by the late Dr. P. P. Carpenter and Mr. Whiteaves. I have had the benefit, in the case of several critical species, of the advice of the late Mr. J. G. Jeffreys, the late Mr. R. MacAndrew of London and Mr. Dall of Washington. I am also indebted to Mr. G. S. Brady for determining the Ostracoda, to the Rev. H. W. Crosskey for opportunities of comparing specimens with those of the Clyde Beds, and to Prof. T. R. Jones and Dr. Parker and Mr. G. M. Dawson for help with the Foraminifera. Other names of contributors of information will be mentioned in the notes and in the lists of fossils.

CHAPTER II.

THE SUCCESSION OF DEPOSITS.

Under this heading I propose to mention, in a general manner, the actual succession of deposits with which we have to deal, and to consider what is known of the probable conditions of their accumulation and the geographical and climatal features of the period.

The deposits which we have to consider are superficial or overlying beds of boulder clay or till, laminated clay, sand and gravel, widely distributed over the northern part of the American continent, and accompanied with ridges of detritus (kaims, eskers, etc.), and with travelled stones and boulders or erratics. They may be arranged as in the table on the following page, which refers to three important and widely separated regions.* It will be observed that in each of these there is a general correspondence in the arrangement of the deposits, and that these may be regarded as comprising a lower and upper boulder formation separated by an intermediate bed or set of beds evidencing a less amount of ice action. This intermediate formation corresponds with that often named "Inter-glacial." A similar order is observed in other parts of America, and also in Western Europe.

---

* Pleistocene Flora of Canada, Bul. Geol. Soc. of America, Vol. I.

COMPARATIVE TABLE.

Montreal and lower St. Lawrence.	North shore of Lake Ontario.	Belly river, North-west Territory.
J. WM. DAWSON.	J. G. HINDE.	G. M. DAWSON.
I.	I.	I.
Surface soil, post-glacial alluvia & peat.	Surface soil, strati-fied sand and gravel.	Surface soil and prairie alluvium.
II.	II.	II.
Surface boulders, Saxicava sand and gravel. Boulders in and below sand.	Boulders, sand, etc. Laminated clay. Upper boulder deposit.	Upper boulder clay.
III.	III.	III.
Upper Leda clay, marine shells and drift plants. Lower Leda clay, marine shells and drift plants.	Stratified sand and clay, with fresh-water shells and plants.	Gray sand with iron-stone nodules. Brownish sandy clay. Carbonaceous layers and peat. Gray sand and ironstone.
IV.	IV.	IV.
Lower boulder clay or till. Many native and some travelled boulders. A few marine shells of arctic species.	Lower boulder clay or till. Native and travelled boulders.	Lower boulder clay. Many travelled boulders.
V.	V.	V.
Palæozoic rocks, often striated.	Palæozoic rocks, often striated.	Cretaceous beds.

(The bracket to the left spanning sections II–V is labelled: Pleistocene.)

Taking a somewhat more general view, the whole pleistocene deposits of eastern Canada may be tabulated in descending order as follows :—

CANADIAN PLEISTOCENE.

(a) Post-glacial deposits, river alluvia and gravels, peaty deposits, lake bottoms, etc. } Remains of *Mastodon* and *Elphas*, modern fresh-water shells.

(b) Saxicava sand and gravel, often with numerous travelled boulders (upper boulder deposit), probably the same with Algoma sand, etc., of inland districts. } Shallow-water fauna of boreal character, more especially *Saxicava rugosa* and its varieties. Bones of whales, etc.

(c) Upper Leda clay, and probably Saugeen clay * of inland districts; clay and sandy clay, in the lower St. Lawrence, with numerous marine shells. } Holds in eastern Canada a marine fauna identical with that of the northern part of the gulf of St. Lawrence at present; and locally affords remains of a boreal flora.

(d) Lower Leda clay; fine clay, often laminated, and with a few large travelled boulders, probably equivalent to Erie clay† of inland districts. } Holds *Leda* (*Portlandia*) *arctica* and sometimes *Tellina groenlandica;* and seems to have been deposited in very cold and ice-laden water.

(e) Lower stratified sands and gravels (Syrtensian deposits of Matthew).‡ } These represent land surfaces and sea and coast areas immediately posterior to the boulder clay.

(f) Boulder clay or till; hard clay, or unstratified sand, with boulders, local and travelled, and stones often striated and polished. } In the lower St. Lawrence region holds a few marine shells of arctic species. Farther inland is non-fossilferous, but has usually the chemical characters of a marine deposit.

NOTE.—With reference to this table, I wish it to be distinctly understood that it covers the whole pleistocene deposits as known in Canada, and that division (f) corresponds to the older boulder clay and (b) to the upper boulder deposit, which is the more extensively spread of the two.

* Geology of Canada, 1862.

† Supplement to Acadian Geology, 1878. Notes on Post-pliocene of Canada: Canadian Naturalist, Vol. VI., 1871.

‡ In the province of Quebec beds of this kind in some places underlie the boulder clay.

The lower boulder clay (*f*) is often a true and very hard till, resting usually on intensely glaciated rock-surfaces, and filled with stones and boulders. Where very thick, it can be seen to have a rude stratification. Even when destitute of marine fossils, it shows its submarine accumulation by the unoxidized and unweathered condition of its materials. The striæ beneath it, and the direction of transport of its boulders, show a general movement from N.E. to S.W., up the St. Lawrence valley from the Atlantic. Connected with it, and apparently of the same age, are evidences of great local glaciers descending into the valley from the Laurentian highlands. The boulder clay of the basins of the great lakes, and of the western plains, as well as that of the Missouri Côteau, seems to be of similar character. The basins of the lakes are parts of older valleys dammed up with Pleistocene debris.* The Missouri Côteau and its extensions, probably the greatest " moraine " in the world, and the "terminal moraine" of the great continental glacier of some American geologists, appears to me to be the deposit at the margin of a sea laden with vast fields of floating ice.†

The lower Leda clay (*d*) seems in all respects similar to the deposits now forming under the ice in Baffin's bay and the Spitzbergen sea. The upper Leda clay represents a considerable amelioration of climate, its fauna being so similar to that of the gulf of St. Lawrence at present, that I have dredged in a living state nearly all the species it contains, off the coasts on which it occurs.

---

* Newberry, Reports on Ohio ; Hunt, Canadian Reports ; Spencer, Ancient Outlet of lake Erie, Ann. Phil. Society, 1881.

† Report on 49th Parallel. G. M. Dawson, Paper on Superficial Deposits of the Plains in the Journal of London Geological Society.

Land plants found in the beds holding these marine shells are of species still living on the north shore of the St. Lawrence, and show that there were in certain portions of this period considerable land surfaces clothed with vegetation. The upper Leda clay is probably contemporaneous with the so-called inter-glacial deposits holding plants and insects discovered by Hinde on the shores of lake Ontario.[*] On the Ottawa it contains land plants of modern Canadian species, insects and feathers of birds, intermixed with skeletons of Capelin (*Mallotus*) and shells living in the gulf of St. Lawrence.

The changes of level in the course of the deposition of the Leda clays must have been very great; fossiliferous marine deposits of this age being found at a height of at least 600 feet, and sea-beaches at a much greater elevation, while at other times there must have been large land areas and even fresh-water lakes. Littoral gravels and sands of this period may also be undistinguishable, except by their greater elevation, from those of the Saxicava sand. I have described the bones of a large whale (*Megaptera longimana*) from gravel north of the outlet of lake Ontario and 420 feet above the level of the sea, which is not improbably contemporaneous with the Leda clay of lower levels, and much higher than deposits near lake Ontario regarded as of lacustrine origin.[†] These

---

[*] Proceedings of Canadian Institute, 1877. Dr. Hinde in this paper incorrectly states that the Leda clay belongs to the "close of the glacial period," and that boulder-drift is not found above it. In truth, as Admiral Bayfield, Sir Charles Lyell, and the writer have shown, boulder-drift is still in progress in the gulf and river St. Lawrence, though in a more limited area than in the pleistocene period; but any considerable subsidence of the land might enable it to resume its former extension.

[†] Canadian Naturalist, Vol. X., No. 7.

changes of the relative levels of sea and land must be taken into account in explaining the distribution of marine clays and sands, boulder deposits, etc., which are often regarded with reference to the present levels of the country, or as contemporaneous deposits without regard to their elevation, a method certain to lead to inaccurate conclusions.

The Saxicava sand (*b*) indicates shallow-water conditions with much driftage of boulders, and probably glaciers on the mountains. It constitutes in many districts a second boulder formation, and possibly implies a somewhat more severe or at least more extreme climate than that of the upper Leda clay. Terraces along the coast mark the successive stages of elevation of the land in and after this period. There is also evidence of a greater elevation of the land succeeding the time of the Saxicava sand, and preceding the modern era.*

It is well known that very diverse theoretical views exist among geologists as to the origin of the deposits above referred to. The conclusions which have been forced upon the writer by detailed studies extending over the last forty years, are that in Canada the condition of most extreme glaciation was one of partial submergence, in which the valleys were occupied by a sea laden with heavy field-ice continuing throughout the summer, while the hills remaining above water were occupied with glaciers, and that these conditions varied in their distribution with the varying levels of the land, giving rise to great local diversities, as well as to changes of climate. There seems to be within the limits of Canada no good evidence of a general covering of the land with a thick

* Supplement to Acadian Geology, 3rd edition, pp. 14, et seq.

mantle of ice, though there must at certain periods have been very extensive local glaciers on the Appalachians, the Laurentian axis and the mountainous regions of the west.* The two latter have been named by Dr. G. M. Dawson the "Laurentide" and "Cordilleran" glaciers respectively. The former may be named the "Appalachian" glacier, and these three must have been the principal sources of land ice in the height of the glacial age, when large portions of the plains and valleys must have been submerged. It does not, indeed, seem possible that, under any conceivable meteorological conditions, an area so extensive as that of Canada, if existing as a land surface, should receive, except on its oceanic margins, a sufficient amount of precipitation to produce a continental glacier.†

In the great Cordilleran ranges of the north-west the changes evidenced in the east occurred in an exaggerated form. The general character and probable complexity of these changes may be seen from the following provisional table taken from Dr. G. M. Dawson, and the evidence for which will be found in his memoir on the "Physiographical Geology of the Rocky Mountain Region of Canada," ‡ already referred to.

---

* G. M. Dawson, Reports on British Columbia, and Superficial Geology of British Columbia, Journal Geol. Society, 1878. Memoir on Rocky Mountains, Trans. R. S. C., 1890.

† The term "modified drift," sometimes used for the upper pleistocene deposits, is objectionable. The gravels and sands of the Saxicava sand are no more "modified" representatives of the lower beds than a carboniferous sandstone or conglomerate is a modification of underlying strata. The term has no proper significance, unless it could be shown that the boulder clay is a deposit formed on land and subsequently modified by aqueous action.

‡ Trans. R. S., Canada, 1890.

4

SCHEME OF CORRELATION OF THE PHENOMENA OF THE GLACIAL
PERIOD IN THE CORDILLERAN REGION AND THE REGION OF
THE GREAT PLAINS (IN ASCENDING ORDER).

Cordilleran Region.	*Region of the Great Plains.*
Cordilleran zone at a high elevation. Period of most severe glaciation and maximum development of the great Cordilleran glacier.	Correlative subsidence and submergence of the great plains, with possible contemporaneous increased elevation of the Laurentian axis and maximum development of ice upon it. Deposition of the lower boulder clay of the plains.
Gradual subsidence of the Cordilleran region and decay of the great glacier, with deposition of the boulder clay of the Interior Plateau and the Yukon Basin, of the lower boulder clay of the littoral, and also at a later stage (and with greater submergence) of the inter-glacial silty beds of the same region.	Correlative elevation of the western part of the great plains, which was probably more or less irregular, and led to the production of extensive lakes, in which inter-glacial deposits, including peat, were formed.
Re-elevation of the Cordilleran region to a level probably as high as or somewhat higher than the present. Maximum of second period of glaciation.	Correlative subsidence of the plains, which (at least in the western part of the region) exceeded the first subsidence and extended submergence to the base of the Rocky Mountains near the forty-ninth parallel. Formation of second boulder clay, and (at a later stage) dispersion of large erratics.
Partial subsidence of the Cordilleran region to a level about 2,500 feet lower than the present. Long stage of stability, during which the white silts were laid down. Glaciers of the second period considerably reduced. Upper boulder clay of the coast	Correlative elevation of the plains, or at least of their western portion, resulting in a condition of equilibrium as between the plains and the Cordillera, their *relative* levels becoming nearly as at present. Probable formation of the Missouri Côteau along a

Cordilleran Region.	*Region of the Great Plains.*
probably formed at this time, though perhaps in part during the second maximum of glaciation.	shore-line during this period of rest.
Renewed elevation of the Cordilleran region with one well marked pause, during which the littoral stood about 200 feet lower than at present. Glaciers much reduced and diminishing, in consequence of general amelioration of climate toward the close of the glacial period.	Simultaneous elevation of the great plains to about their present level, with final exclusion of waters in connection with the sea. Lake Agassiz formed and eventually drained toward the close of this period. This simultaneous movement in elevation of both great areas may probably be connected with the more general northern elevation of land at the close of the glacial period.

The tendency of recent observations has been to show that the Pliocene and older subdivisions of the Tertiary covered each of them a much longer time than the Pleistocene, and that the close of the latter approaches more nearly to the modern or recent time than had previously been supposed. To these points we shall have occasion to refer in the sequel.

It may be proper here to indicate the general nomenclature which will be followed. When the whole geological series is divided into Primary, Secondary and Tertiary, the deposits to which this paper relates are usually named Post-tertiary or Quaternary. These terms are, in my judgment, unfortunate and misleading. If we take the relations of fossils as our guide, then, as Pictet has well remarked, whether we regard the land or the sea animals, there is no decided break between the newer Pliocene and the Pleistocene, the changes not being greater than those between the Pliocene and the older Tertiary ages. There is, therefore, no such thing in nature as a Quaternary time distinct from the Tertiary, as the Tertiary is distinct from the Secondary. Where therefore the terms Primary, Secondary and Tertiary are used, the latter should include the whole time from the Eocene to the Modern inclusive, unless indeed the advent of man be considered an event of sufficient geological importance to warrant a separation of the Modern from the Tertiary period. When the terms Palæozoic, Mesozoic and Kainozoic or Neozoic are used, then the two latter terms cover perfectly the Pleistocene as well as the Eocene, Miocene and Pliocene. I would therefore include the Pleistocene in the Neozoic or Tertiary period, and define it to be that geological age which is included between the Pliocene and the Modern. From the former it is separated by the advent of the cold or glacial* period, and the accompanying subsidence of the land, as well as by the disappearance of many species of animals and plants. From the latter it is separated by the extinction of many mammalian forms, and by the introduction of man, and of the present levels and climatal conditions of the continents.

LATER KAINOZOIC OR TERTIARY PERIOD.

(In Ascending Order.)

NEWER PLIOCENE.—A continental period of long duration, in which the land was more elevated than at present, and very extensive erosion of deep river valleys occurred.

PLEISTOCENE.—Covering three sub-divisions :—

    (a) *Early Pleistocene:* Irregular elevation and depression of the continents, with cold climate and great local glaciers.

---

* I use the term "glacial" in this paper in its general sense, as including the action of floating ice as well as of land ice.

(b) Mid-Pleistocene: Submergence of coasts and re-elevation of interior plateaus with milder climate.—Inter-glacial period.

(c) Later Pleistocene: Submergence of plains, and general ice drift, with local glaciers in mountains.

EARLY MODERN OR POST-GLACIAL.—Second continental period, in which the land regains almost all the extension of the Pliocene time. Age of the Mammoth and Mastodon and of Palæocosmic man.—Post-glacial Fauna.

MODERN OR RECENT.—Submergence of short duration, terminating the age of Palæocosmic man. Re-elevation of continents to present levels. Modern races of men and Modern Fauna.

Let us now consider the several members of the Pleistocene more in detail, especially in those regions in which they have been studied by the author.

GENERAL DESCRIPTION OF PLEISTOCENE DEPOSITS.

1.—*The Lower Boulder-Clay.*

Throughout a great part of Canada there is over all the lower levels a true "Till," consisting of hard gray clay, filled with stones and thickly packed with boulders. In many places, however, the clay becomes sandy, and in some portions of the upper carboniferous and triassic areas, the paste is an incoherent sand. The mass is usually destitute of any stratification or subordinate lamination; but sometimes in thick beds horizontal lines of different texture or colour can be perceived, and occasionally the clay intervening between the stones becomes laminated, or at least shows such a structure when disintegrated by frost. The boulder-clay usually rests directly on striated rock-surfaces; but I have observed in Cape Breton a peaty or brown coal deposit, with branches of coniferous trees, which underlies it, and in other places there are deposits of rolled gravel under the boulder-clay. At the Glen brick-work, near Montreal, a peculiar modified boulder-clay occurs, consisting of very irregularly bedded sand and gravel, with many large boulders, and only thin layers of clay.

The stones of the boulder-clay are often scratched, and ground into those peculiar wedge-shapes so characteristic of ice-worked stones. Very abundant examples of this occur at Montreal and in its vicinity.

At Isle Verte, Rivière du Loup, Murray Bay, Quebec, St. Nicholas, Little Metis, etc., the boulder-clay is fossiliferous, containing especially *Leda glacialis*, and often having boulders and large stones covered with *Balanus Hameri* and with Bryozoa, evidencing that they have for some time quietly reposed in the sea bottom before being buried in the clay. This is indeed the usual condition of the boulder-clay in the lower part of the St. Lawrence river.* Further up, in the vicinity of Montreal, it has not been observed to contain fossils, but it presents equally unequivocal evidence of sub-aqueous origin in the low state of oxidation of the iron in the blue clay, which becomes brown when exposed to the weather, and in the brightness of the iron pyrites contained in some of the glaciated stones, as well as in the presence of rounded and

---

* Upham admits (Proc. Brit. Nlt. Soy., 1888) that sea shells exist in the boulder-clay of Massachusetts; but his explanation that they have been pushed up by glaciers is quite inadmissible, more especially as they are not of more boreal types than those of Massachusetts bay at present.

glaciated lumps of Utica shale and other soft rocks, which become disintegrated at once when exposed to weathering.

The true boulder-clay is in all ordinary cases the oldest member of the Pleistocene deposits, and it is not possible to divide it into distinct boulder-clays of different ages, superimposed on one another. It may be observed, however, that in so far as the boulder-clay is a marine deposit, that which occurs at lower levels is in all probability newer than that which occurs at higher levels. It is also to be observed that boulders with layers of stones occasionally occur in the Leda clay; and that the superficial sands and gravels sometimes contain large boulders, and even constitute an upper or newer boulder formation; but these appearances are not usually sufficiently important to cause any experienced observer to mistake such overlying deposits for the lower boulder-clay. They belong to the second or newer part of the period.

In some localities the stones in the boulder-clay are almost exclusively those of the neighbouring rock formations, and this is especially the case at the base of cliffs or prominent outcrops, whence a large quantity of material would be easily derived. In other cases, though less frequently, material travelled from a distance largely predominates. Throughout the valley of the Lower St. Lawrence, the gneiss and other hard metamorphic rocks of the Laurentian hills to the north-east are very abundant, and in boulders of large size and much rounded. Occasional instances also occur where large boulders have been transported to the northwards; but these are comparatively rare, except in the second or upper drift. I have mentioned some examples of this in "Acadian Geology," p. 61. Similar instances are mentioned in the "Geology of Canada," p. 893.

Though the boulder-clay often presents a somewhat widely extended and uniform sheet, yet it may be stated to fill up all small valleys and depressions, to be confined chiefly to the lower grounds, and to be thin or absent on ridges and rising grounds. The boulders which it contains are also by no means uniformly dispersed. Where it is cut through by rivers, or denuded by the action of the sea, ridges of boulders often appear to be included in it. Those on the Ottawa referred to in the "Geology of Canada," p. 895, are very good illustrations, and I have observed the same fact on the Lower St. Lawrence and on the coast of Nova Scotia. It is also observable that these lines and groups of boulders are often not of local material, but of rocks from distant localities, and that a number of the same kind seem often to have been deposited together in one group.

Loose boulders are often found upon the surface, and sometimes in great numbers. In some instances these may represent beds of boulder-clay removed by denudation. In other cases they may have been derived from the overlying members of the formation, or may have been deposited on the surface in the later Pleistocene subsidence, without any covering of clay or gravel. In "Acadian Geology," p. 64, I have illustrated the manner in which large stones, sometimes eight feet or more in diameter, are moved by the coast ice and sometimes deposited on the surface of soft mud, and I have had occasion to verify the observations of the same kind made by Admiral Bayfield, and quoted by Sir C. Lyell in the "Principles of Geology." Lastly, on certain high grounds there are large loose boulders, which have probably been moved to their present positions by means of land ice or glaciers.

The boulder-clay locally presents, as above stated, indications of successive layers, and it occasionally contains
surfaces on which lie large boulders striated and polished
on the upper surface, in the manner of the pavements of
boulders described by Miller, as occurring in the Till of
Scotland. These appearances are, however, rare, and few
opportunities occur for observing them.

A very general and important appearance is the polishing and striation of the underlying rocks usually to be
observed under the boulder-clay, and which is undoubtedly
of the same character with that observed under Alpine
glaciers. This continental striation or grooving is obviously the effect of the action of ice, and its direction
marks the course in which the abrading agent travelled.
This direction has been ascertained by the Canadian and
United States surveys, and by local observers, over a
large part of America, and it presents some broad features
well deserving attention. A valuable table of the direction of this striation is given in the "Geology of Canada,"
which I may take as a basis for my remarks, adding to it
a few local observations of my own.* The table embraces
one hundred and forty-five observations, extending along
the valleys of the St. Lawrence and the Ottawa and the
borders of the great lakes. In all of these the direction
is south, with an inclination to the west and east, or to
state the case more precisely, there are two sets of striae,
a south-west set and a south-east set. In the table
eighty-four are westward of south and fifty-eight are
eastward of south, three being due south. It further

---

* See also, for the western districts, Whittlesey's Memoir in the
Smithsonian Contributions, and Newberry's Report on Ohio ; Papers
by Dr. G. M. Dawson on the Plains of N. W. Canada in Journal of
Geol. Soy. of London and Trans. R. S. Canada.

appears, when we mark the localities on the map, that in
the valley of the St. Lawrence and the rising grounds
bounding it, the prevailing course is south-west, and this
is also the prevalent direction in western New York, and
behind the great Laurentide chain on the north side of
lake Huron. Crossing this striation nearly at right
angles, is a second set, which occurs in the neck of land
between Georgian bay and lake Ontario, in the valley of
the Ottawa and in the hilly districts of the Eastern
Townships of the province of Quebec, where it is connected with a similar striation which is prevalent in the
valleys of lake Champlain and the Connecticut river and
elsewhere in New England. In New England this
striation is said to have been observed on hills 4,800 feet
high, as for example on Mansfield mountain, where,
according to Hitchcock, there are striae bearing S. 30° E.
at an elevation of 4,848 feet. In Nova Scotia and New
Brunswick, as in New England, the prevailing direction
is south-eastward, though there are also south-west and
south striation, and a few cases where the direction is
nearly east and west. Recent observations lead to the
belief that in eastern Canada the south-west and north-
east striation is general on the lower grounds. The
south-east and north-west striation belongs more to the
higher grounds, and in some cases represents ice-flow *in
two directions*, to the north-west and south-east of the
ridges of high land.

It is obvious that such striation must have resulted
from the action of a solid mass or masses of ice bearing
for a long time on the surface, and abrading it by means
of stones and sand. It is further obvious that the
different sets of striation could scarcely have been produced at the same time in any one locality, especially

when, as is not infrequent, we have two sets nearly at right angles to each other, in the same locality. Hence it becomes an important question to ascertain the relative ages of the striation, and also the direction in which the abrading force moved.

Taking the valley of the St. Lawrence in the first instance, the crag-and-tail forms of the isolated hills of trap, like the Montreal mountain, with abrupt escarpments to the north-east and slopes of debris to the south-west, the quantity of boulders carried from them far to the south-west, and the prevailing striation in the same direction, all point to a general movement of detritus up the St. Lawrence valley to the south-west. Further, in some cases the striae themselves show the direction of the abrading force. For example, in a fine exposure recently made at the Mile-end quarries, near Montreal, the polished and grooved surface of the limestone shows four sets of striae. The principal ones have the direction of S. 68° W. and S. 60° W. respectively, and the second of these sets is the stronger and coarser, and sometimes obliterates the first. The two other sets are comparatively few and feeble striae, one set running nearly N. and S., and the other N.W. and S.E. These last are probably newer than the two first sets. Now, with regard to the direction of the principal sets of striae, this at the locality in question was rendered very manifest by the occurrence of certain trap dykes crossing the limestone at right angles to the striae. The force, whatever it was, had impinged on these dykes from the N.E., and their S.W. side had protected the softer limestone. The locality is to the N.E. of the mass of trap constituting the Montreal mountain, and the movement must have been up the St. Lawrence valley from the N.E., and toward the mountain, but at this

particular place the striae point west of its mass. This, I
have no hesitation in saying, is the dominant direction in
the St. Lawrence valley, and it certainly points to the
action of the arctic current passing up the valley in a
period of submergence. Further, it is the boulder-clay
connected with this S.W. striation that has hitherto
proved most rich in marine shells.

If, however, we pass from the St. Lawrence valley up
the valleys which open into it from the north, as for
example the gorge of the Saguenay, the Murray Bay river,
or the Ottawa river, we at once find a striation nearly at
right angles to the former, or pointing to the south-east.

At the mouth of the Saguenay, near Moulin Bode, are
striae and grooves on a magnificent scale, some of the
latter being ten feet wide and four feet deep, cut into
hard gneiss. Their course is N. 10° W. to N. 20° W.
magnetic, or N. 30° to 40° W. when referred to the true
meridian. In the same region, on hills 300 feet high, are
roches moutonnees with their smoothest faces pointing in
the same direction, or to the north-west. This direction
is that of the valley or gorge of the Saguenay, which
enters nearly at right angles the valley of the St.
Lawrence. At the mouth of the Saguenay the Lark
Shoals constitute a mass of debris and boulders, both
inside and outside of which is very deep water; and many
of the fragments of stone on these shoals must have been
carried down the Saguenay more than fifty miles.

In like manner at Murray bay there are striae on the
Silurian limestones near Point au Pique, which run about
N. 45° W., but these are crossed by another set having a
course S. 30° W., so that we have here two sets of
markings, the one pointing upwards along the deep valley
of Murray Bay river to the Laurentide hills inland, the

other following the general trend of the St. Lawrence valley. The boulder-clay which rests on these striated surfaces is a dark-coloured till, full of Laurentian boulders, and holding *Leda glacialis*, and also Bryozoa clinging to some of the boulders. In ascending the Murray Bay river, we find these boulder-beds surmounted by very thick stratified clays, with marine shells, which extend upward to an elevation of about 800 feet, when they give place to loose boulders and unstratified drift. About this elevation, the laminated clays meet a ridge of drift like a moraine, crossing the valley, which forms the barrier of a small lake, Petite Lac, and a second similar barrier separates this from Grand Lac. If the valley of Murray Bay river was occupied with a glacier descending from the Laurentian hills inland, which are probably here 3,000 to 4,000 feet high, this glacier or large detached masses pushed from its foot, must have at one time extended quite to the border of the St. Lawrence, and at another must have terminated at the borders of the two lakes above mentioned.

On a still larger scale the N.W. and S.E. striation appears in the valley of the Ottawa, and farther west between the head of lake Ontario and lake Huron, in the valleys descending from the Laurentian plateau. Here it may be ascribed in part to general ice-laden currents from the north-west, and in part to portions of the great Laurentide glacier.

A most important observation bearing on this subject appears in the Report of Mr. R. Bell, in the region of lake Nipigon, north of lake Superior. He observed there the prevailing south-west striation, but with a more westerly trend than usual. Crossing this, however, there was a southerly and S.E. set of striae which were observed

to be older than the south-west striae. In some other
parts of Canada these striae seem to be newer than the
others, but there would be nothing improbable in their
occurring both at the beginning and end of the boulder-
clay period.

In summing up this subject, I think it may be affirmed
that when the striation and transfer of materials have
obviously been from N.E. to S.W., in the direction of the
arctic current, and more especially when marine remains
occur in the drift, we may infer that floating ice and
marine currents have been the efficient agents. Where
the striation has a local character, depending upon existing mountains and valleys, we may on the other hand
infer the action of land ice. For many minor effects of
striation, and of heaping up of moraine-like ridges, we
may refer to the presence of lake or coast ice as the land
was rising or subsiding. This we now see producing such
effects, and I think it has not been sufficiently taken into
the account.

As to the St. Lawrence valley, it is evident that its
condition during the deposit of the boulder-clay must
have been that of a part of a wide sound or inland sea
extending across the continent, and that local glaciers
may have descended into it from the high lands on the
north, and on the south which may have been relatively
higher than at present. During this state of the valley

Fig. 1.—Travelled Boulder on Glaciated Rock. (After Dr. G. M. Dawson.)

great quantities of boulders were brought down into it, especially from the Laurentide hills, and were drifted along the valley, principally to the south-west. Extensive erosion also took place by the combined action of frost, rain, melting snows, and the arctic current and the waves, and thus was furnished the finer material of the boulder-clay. On the south shore of the St. Lawrence, the Notre Dame mountains, stretching out towards cape Gaspé, afford indications of local glaciation, and Mr. R. Chalmers has shown that the movement of ice from this elevated region has been both south toward the baie des Chaleurs, and north toward the St. Lawrence.* I have myself seen ample evidence in large travelled boulders of Silurian limestone on the south shore of the St. Lawrence, of drift from the hills on the south intermixed with that from the Laurentians on the north. Similar facts have been observed by Ells and Low in the hills of the Eastern Townships of the province of Quebec.

It is further to be observed that oscillations of land must be taken into account in explaining these phenomena. Elevations increasing the height and area of land might increase the space occupied by snow and land ice. Depressions, on the other hand, would bring larger areas under the influence of water-borne ice and marine deposits, and these might take place either in a shallow sea loaded with field and coast ice, or in deeper water in which large icebergs might float or ground. The effects would be the greater if, as Dr. G. M. Dawson has shown in the case of the Cordilleran chain, there was unequal elevation causing contemporaneous depression of the

* On the Glaciation and Pleistocene subsidence of northern New Brunswick and south-eastern Quebec. Trans. R.S.C., 1886.

plains and elevation of mountains. There is reason to believe that such alternations were not infrequent in the Pleistocene, and that their occurrence will explain many of the complexities of these deposits.

If we adopt for the more general deposits the hypothesis of floating ice, we must be prepared to consider in connection with this subject a subsidence so great as to place at one period all but the highest parts of the Laurentides and Appalachians under water. In this case a vast volume of arctic ice and water would pour over the country of the great lakes to the S.W., while any obstruction occurring to the south would throw lateral currents over the Appalachians to the eastward.

It is evident from the descriptions of Smith, Geikie, Jameson, Crosskey, and others, that the boulder-clay of Scotland and Scandinavia corresponds precisely in character with that of Canada, and there, as in America, the theory of a continental glacier has been resorted to for its explanation. The objections to this hypothesis are very ably stated by Mr. Milne Home in a paper on the "Boulder-clay of Europe," in the Transactions of the Royal Society of Edinburgh, 1869.

To this period and these causes must also be assigned the excavation of the basins of the great American lakes. These have been cut out of the softer members of the Silurian and Devonian Formations; but the mode of this excavation has been regarded as very mysterious; and, like other mysteries, has been referred to glaciers. Its real cause was obviously river and atmospheric erosion in the Pliocene period, supplemented by the flowing of cold ocean currents over the American land during its submergence.* The lake-basins are thus of the same nature

---

* See Chapter III.

with the deep hollows extending outward from the river mouths of the American coasts under the ocean, or perhaps they are like those intervening between the banks cast up by the arctic currents on the present American coast, and like those deep channels of the arctic current in the Atlantic recently explored by Dr. Carpenter. Their arrangements geographically, as well as their geological relations, correspond with such views.

The former consideration with regard to the great lakes deserves especial notice. Drs. Hunt, Newberry, and Spencer have collected many facts to show that the lake basins are connected with one another and with the sea by deep channels now filled up with drift-deposits. It is therefore certain that much of the erosion of these basins may have occurred before the advent of the glacial period, in the Pliocene age, when the American continent was at a higher level than at present. Dr. Newberry has given in the Report in the Geology of Ohio, a large collection of facts ascertained by boring or otherwise, which go far to show that were the old channels cleared of drift and the continent slightly elevated, the great lakes would be drained into each other and into the ocean by the valleys of the Hudson and the Mississippi, without any rock cutting, and if the barrier of the Thousand Islands were then somewhat higher, the St. Lawrence valley might have been cut off from the basin of the great lakes. Spencer has, however, shown, on the evidence of differential elevation, that a portion at least of the drainage of the Pliocene lake country may have found its way down the present course of the St. Lawrence valley.

The latter cause, namely, the possible eroding action of ocean currents, is one more difficult to estimate, yet should not be neglected by geologists. I thus referred to it in 1864.[*]

"Our American lake-basins are cut out deeply in the softer strata. Running water on the land could not have done this under the present geographical conditions, though it could effect it with a higher level and better drainage; nor could the result be effected by ocean breakers, though the levelling power of these is enormous. Glaciers could not have effected it; for even if the climatal conditions for these were admitted, there is no height of land to give them momentum. But if we suppose the land submerged so that the arctic current, flowing from the north-east, should pour over the Laurentian rocks on the north side of lake Superior and lake Huron, it would necessarily cut out of the softer Silurian strata just such basins, drifting their materials to the south-west. At the same time, the lower strata of the current would be powerfully determined through the strait between the Adirondac and Laurentide hills, and running over the ridge of hard rock which connects them at the Thousand Islands, would cut out the long basin of lake Ontario, heaping up at the same time in the lee of the Laurentian ridge, the great mass of boulder-clay which intervenes between lake Ontario and Georgian bay. Lake Erie may have been cut by the flow of the upper layers of water over the Middle Silurian escarpment; and lake Michigan, though less closely connected with the direction of the current, is, like the others, due to the action of a continuous eroding force on rocks of unequal hardness.

---

[*] Presidential Address to Nat. Hist. Soc. of Montreal, Canadian Naturalist, 1864.

" The predominant south-west striation, and the cutting
of the upper lakes, demand an outlet to the west for the
arctic current. But both during depression and elevation
of the land, there must have been a time when this outlet
was obstructed, and when the lower levels of New York,
New England, and Canada were still under water. Then
the valley of the Ottawa, that of the Mohawk, and the
low country between lakes Ontario and Huron, and the
valleys of lake Champlain and the Connecticut, would be
straits or arms of the sea, and the current, obstructed in
its direct flow, would set principally along these, and act
on the rocks in north and south, and north-west and
south-east directions. To this portion of the process we
may attribute much of the north-west and south-east
striation. It is true that this view does not account for
the south-east striae observed on some high peaks in New
England ; but it must be observed that even at the time
of greatest depression, the arctic current would cling to
the northern land, or be thrown so rapidly to the west
that its direct action might not reach such summits.
There were also extensive local glaciers in these mountains, whose work must be separated from that of the
sea-drift.

" I conclude these remarks with a mere reference to
the alleged prevalence of lake-basins and fiords in high
northern latitudes, as connected with glacial action. In
reasoning on this, it seems to be overlooked that the prevalence of disturbed and metamorphic rocks over wide
areas in the north is one element in the matter, and that
in the Pliocene age the greater elevation of the land must
have caused deeper fluviatile erosion. Further, the fiords
on coasts, like the deep lateral valleys of mountains, are
often evidences of the action of the waves rather than of

that of ice. I am sure that this is the case with many of
the indentations of the coast of Nova Scotia, which are
cut into the softer and more shattered bands of rock,
and show, in raised beaches and gravel ridges like those
of the present coast, the levels of the sea at the time of
their formation."

To the period of the boulder-clay we may refer those
ridges and pavements of boulders imbedded in this clay
or continuous with it, and which testify to the carrying
and packing power of ice. We shall find, however, that
such moraine-like ridges are not confined to this period,
but occur along the sea-margins of the Later Pleistocene,
and are even at this day in process of formation on a
considerable scale along the borders of the St. Lawrence.

2.—The Leda Clay.

This deposit constitutes the subsoil over a large portion
of the great plain of Lower Canada, varying in thickness
from a few feet to 50 or perhaps even 100 feet, and
usually resting on the boulder-clay, into which it sometimes appears to graduate, the material of the Leda clay
being of the same nature with the finer portion of the
paste of the boulder-clay. Its name is derived from the
presence in it of shells of *Leda glacialis*, often to the
exclusion of other fossils, and usually in a perfect state,
with both valves united.

The typical Leda clay in its recent state is usually gray
in colour, unctuous, and slightly calcareous. Some beds,
however, are of a reddish hue; and in thick sections
recently cut, it can be seen to present layers of different
shades and occasional thin sandy bands, as well as layers
studded with small stones. It sometimes holds hard
calcareous concretions, which, as at Green's creek on the

Ottawa, are occasionally richly fossiliferous, but more usually are destitute of fossil remains. When dried, the Leda clay becomes of stony hardness, and when burned, it assumes a brick-red colour. When dried and levigated, it nearly always affords some foraminifera and shells of ostracoids; and in this, as well as in its colour and texture, it closely resembles the blue mud now in process of deposition in the deeper parts of the gulf of St. Lawrence.

The lamination of the Leda clay and its included sand layers, show that it was deposited at intervals, between which intervened spaces when currents carried small quantities of sand over the surface. In these intervals shells as well as sand were washed over the bottom, while ordinarily Leda, Nucula and Astarte burrowed in the clay itself. The layers and patches of stones I attribute to deposit from floating ice, and to the same cause must be attributed the large Laurentian boulders, occasionally though rarely seen imbedded in the clay.

The material of the Leda clay has been derived mainly from the waste of the lower Silurian shales of the Quebec and Utica groups, which occupy a great space in the basin of the gulf and river St. Lawrence. The driftage of this material has been to the south-west, and in that direction it becomes thinner and finer in texture. The supply of this mud, under the action of the waves, of streams, of the arctic currents and tidal currents, and floating ice, must have been constant, as it now is in the gulf and river St. Lawrence. It would be increased by the melting of the snows in spring and by any oscillations of level, and it is probably in these ways that we should account for the alternations of layers in the deposit.

The modern deposit in the gulf of St. Lawrence, the chemical characters and coloration of which I explained

many years ago,* shows us that the Leda clay, when in
suspension, was probably reddish or brown mud tinted
with peroxide of iron, like that which we now see in the
lower St. Lawrence; but like the modern mud, so soon as
deposited in the bottom, the ferruginous colouring matter
would, in ordinary circumstances, be deoxidised by organic
substances, and reduced to the condition of sulphide or
carbonate of the protoxide. This colour, owing to its
impermeability, it still retains when elevated out of the
sea; but when heated in presence of air, or exposed for
some time at the surface, it becomes red or brown. The
occasional layers of reddish Leda clay indicate places or
times when the supply of organic matter was insufficient
to deoxidise the iron present in the mass.

The greater part of the Leda clay was probably
deposited in water from twenty to one hundred fathoms
in depth, corresponding to the ordinary depths of the
present gulf of St. Lawrence; and as we shall find, this
view is confirmed by the prevalent fossils contained in it,
more especially the Foraminifera. The most abundant
of these in the Leda clay is *Polystomella striatopunctata*
var. *arctica*, which is now most abundant at about twenty-
five or thirty fathoms. Since, however, the shallow-water
marine Post-pliocene beds extend upwards in some places
to a height of six hundred feet on the hills on the north
side of the St. Lawrence, it is probable that deposits of
Leda clay contemporaneous with these high-level marine
beds were formed in the lower parts of the plain at depths
exceeding one hundred fathoms.

The western limits of the Leda clay appear to occur
where the Laurentian ridge of the Thousand Islands

---

* Journal of Geological Society of London, Vol. V., pp. 25 to 30.

crosses the St. Lawrence, and where the same ancient rocks cross the Ottawa; and in general the Leda clay may be said to be limited to the lower Silurian plain, and not to mount up the Laurentian and metamorphic hills bounding it. Since, however, the level of the water, as indicated by the terraces in Lower Canada, and by the probable depth at which the Leda clay was deposited, would carry the sea level far beyond the limits above indicated, and even to the base of the Niagara escarpment, we must suppose, either—(1) that the supply of this sediment failed toward the west; or (2) that the mud has been removed by denudation or worked over again by the fresh waters so as to lose its marine fossils; or (3) that the relative levels of the western or eastern parts of Canada were different from those at present; or (4) that the water may have been freshened and rendered cold by the influx of melting snow and ice into a landlocked water area or one with a narrow opening. As already stated, there are indications that the first may be an element in the cause. The second is no doubt true of the clays which lie in the immediate vicinity of the lake basins. Dr. Spencer has detailed many observations in favour of the third, more especially in the later glacial and Post-glacial periods.

I believe, however, that much more rigorous investigations of the clays of western Canada are required before we can certainly affirm that none of them hold marine fossils.*

Whittlesey has described the western drift deposits in the Smithsonion Contributions, Vol. XV., and according

---

* It is to be observed that even near the coast the greater part of the thickness of the Leda clay is often unfossiliferous.

to him the boulder-drift is there the upper member of the series. More recently Prof. Newberry has given a summary of the facts in his Report of the Geological Survey of Ohio for 1869. From these sources I condense the following statements:

The lowest member of the western drift, corresponding to the Erie clays of the Canadian Report, is very widely distributed, and fills up the old hollows of the country, in some cases being two hundred feet or more in thickness. Toward the north these clays contain boulders and stones, but do not constitute a true boulder-clay. They rest, however, on the glaciated rock surfaces. They have afforded no fossils except drifted vegetable remains, which appear to occur in an "interglacial" or forest bed between lower and upper boulder-deposits.

Above these clays are sands of variable thickness. They contain beds of gravel, and near the surface teeth of elephants have been found. On the surface are scattered boulders and blocks of northern origin, often of great size, and in some cases transported two hundred miles from their original places. More recent than all these deposits are the "Lake Ridges," marking a former extension of the great lakes.

I believe the Leda clays throughout Canada to constitute in the main one contemporaneous formation. Of course, however, it must be admitted that the deposit at the higher levels may have ceased and been laid dry while it was still going on at lower levels nearer the sea, just as a similar deposit still continues in the gulf of St. Lawrence. On the whole, then, while we regard this as one bed, stratigraphically, we may be prepared to find that in the lower levels the upper layers of it may be somewhat more modern than those portions of the deposit occurring on higher ground and farther from the sea.

Where the Leda clay rests on marine boulder-clay, the change of the deposits implies a diminution of ice-transport relatively to deposition of fine sediment from water; and with this, more favourable circumstances for marine animals. This may have arisen from geographical changes diminishing the supply of ice from local glaciers, or obstructing the access of heavy icebergs from the arctic regions. At the present time, for example, the action of the heaviest bergs is limited to the outer coasts of Labrador and Newfoundland, and a deposit resembling the Leda clay is forming in the gulf of St. Lawrence; but a subsidence which would determine the arctic current and the trains of heavy bergs into the gulf, would bring with it the conditions for the formation of a boulder-clay, more especially if there were glaciers on the Laurentide hills to the north. Where the Leda clay rests on boulder-clay which may be supposed to be of terrestrial origin, subsidence is of course implied; and it is interesting to observe that the conditions thus required are the reverse of each other. In other words, elevation of land or sea bottom might be required to enable Leda clay to take the place of marine boulder-clay, but depression of the land would be necessary to enable Leda clay to replace the moraine of a glacier. I cannot say, however, that I know any case in Canada where I can certainly affirm that this last· change has occurred; though on the north shore of the St. Lawrence there are cases in which the Leda clay rests directly on striated surfaces which might be attributed to glaciers; just as in the west the Erie clay occupies this position.

Deposits referable to the shores of the Leda clay sea,

and the estuaries opening into it from the portions of the
land still above water are not uncommon. Of this nature
are the beds at Pakenham, examined by the late Sheriff
Dickson, and which, as I was informed by him, are
arranged as follows :

	Feet.	Inches.
Sand and surface soil..............about	10	0
Clay....................................	10	0
Fine gray sand (shells of *Valvata*, &c.)....	0	2
Clay.................................	1	0
Gray sand, laminated (*Tellina Greenlandica*)	0	3
Clay...............................	0	8
Light gray sand (*Valvata, Cyclas, Paludina,*		
Planorbis and *Tellina*).................	0	10
Clay.............................	1	2
Brown sand and layers of clay (*Planorbis*		
and *Cyclas*...........................	0	4

The fresh water species are peculiar to this locality,
and the only marine shell is *Tellina Grœnlandica,* a species
now found farther up in our estuaries than most others.

Mr. Dickson informs me that a similar case occurs near
Clarenceville, about four miles from the United States
frontier, and at an elevation of about ten feet above lake
Champlain.   Specimens from this place contain large
shells of *Unio rectus* and *U ventricosus,* the latter with the
valves cohering, and a *Lymnea.*   Intimately mixed with
these in sandy clay are valves of *Tellina Grœnlandica* and
Mya arenaria.

I record these facts, without pledging myself to the
conclusion that these deposits really mark the margins or
river estuaries of the old Pleistocene of Canada, though
they will certainly bear that interpretation.   In farther
connection with these facts, and in relation also to the
question why marine fossils have not been found west of
Kingston, Mr. Dickson informs me that fossil capelin are

found on the Chaudière lake, 183 feet above lake St.
Peters, on the Madawaska 206 feet, and at Fort Coulonge
lake 365 feet above the same level, a very interesting
indication of the gradual recession of the capelin spawn-
ing grounds from this last high elevation to the level of
the more celebrated locality of these fossils at Green's
creek.   Farther, throughout the counties of Renfrew,
Lanark, Carleton and Leeds, the marine deposits rise to an
elevation of 425 feet, or nearly the same with one of the
terraces on Montreal mountain; but while this elevation would, with the present levels of the country, carry
a deep sea to the head of lake Ontario, no marine fossils
appear to have been found on the banks of that lake.
Was the depression of the later Pleistocene period limited
to the country east of lake Ontario, or have the marine
deposits of the upper St. Lawrence hitherto escaped
observation or been removed by denuding agencies ?   The
question awaits further explanations for a satisfactory
answer.

3.—*The Saxicava Sand, and Upper Boulder Deposit.*

When this deposit rests upon the Leda clay, as is not
unfrequently the case, the contact may be of either of two
kinds.   In some instances the surface of the clay has
experienced much denudation, being cut into deep
trenches, and the sand rests abruptly upon it.   In other
cases there is a transition from one deposit to the other,
the clay becoming sandy and gradually passing upwards
into pure sand or fine gravel.   In this last case the lower
part of the sand at its junction with the clay is often
very rich in fossils, showing that after the deposition of
the clay a time of quiescence supervened with favourable
conditions for the existence of marine animals, before the

sand was deposited. It is usually, indeed, in this position that the greater part of the shells of our Post-pliocene beds occur; the Saxicava sand being generally somewhat barren, or containing only a few shallow-water species, while the Leda clay is usually also somewhat scantily supplied with shells, except toward its upper layers. Hence it is somewhat difficult to refer a large part of the shells to either deposit. I have, however, usually regarded the richly fossiliferous deposit as belonging to the Leda clay; and where, as sometimes happens, the clay itself is absent and merely a thin layer rich in fossils separates the Saxicava sand from the boulder-clay, I have regarded this layer as the representative of the Leda clay. Where, on the other hand, the Leda clay is thick and well developed, it admits of sub-division into a *lower Leda clay*, unfossiliferous or with only shells of *Leda glacialis* and *Macoma Grœnlandica*, and an *upper Leda clay*, usually more sandy and holding a rich boreal fauna identical with that of the northern part of the gulf and river St. Lawrence at present.

The Saxicava sand, in typical localities, consists of yellow or brownish quartzose sand, derived probably from the waste of the Potsdam sandstone and Laurentian gneiss, and stratified. It often contains layers of gravel, and sometimes is represented altogether by coarse gravels. It is somewhat irregular in its distribution, forming banks and mounds, partly no doubt in consequence of original irregularities of deposit, and partly from subsequent denudation. In some outlying localities it is liable to be confounded with the modern river sands and gravels. Large travelled boulders often occur in it; but it rarely contains glaciated stones, the stones and pebbles seen in it being usually well rounded.

From the nature of the Saxicava sand, it is obvious that it is for the most part a shallow-water deposit, belonging to the period of emergence of the land; and it must have been originally a marginal and bank deposit, depending much for its distribution on the movement of tides and currents. In some instances, as at Côte des Neiges, near Montreal, and on the terraces on the lower St. Lawrence, it is obviously merely a shore sand and gravel, like that of the modern beach. Ridges of Saxicava sand and gravel have often been mistaken for moraines of glaciers; but they can generally be distinguished by their stratified character and the occasional presence of animal remains, as well as by the water-worn rather than glaciated appearance of their stones and pebbles. In this connection, however, it must be observed that it is not possible to distinguish the high-level beaches and deposits of superficial travelled boulders from the Saxicava sand. In other words, *while the Saxicava sands and gravels may be shallow-water deposits, they must, when at high levels, have been formed on the margins of deep seas.* This is a most important 'fact in connection with the upper or later boulder deposit.

The Saxicava sand sometimes rests on the Leda clay or boulder-clay, and sometimes directly on the rock, and the latter is often striated below this deposit; but in this case there is generally reason to believe that boulder-clay has been removed by denudation. It is to be observed, however, that the typical Saxicava sand and the upper or newer boulder-drift belong to the same period of submergence.

4.—*Terraces and Inland Sea Cliffs, and Kaims.*

These are closely connected with the deposits last mentioned, inasmuch as they have been formed by the

Fig. 2.—Pleistocene terraces at Tadousac bay. Lower terrace, Leda clay ; upper, sand, with drift-stones and boulders on surface.

same recession of the sea which produced the Saxicava sand. At Montreal, where the isolated mass of trap flanked with lower Silurian beds, constituting Mount Royal, forms a great tide-gauge for the recession of the Post-pliocene sea, there are four principal sea margins, with several others less distinctly marked. The lowest of these, at a level of about 120 feet above the level of the sea at lake St. Peter, may be considered to correspond with the general level of the great plain of Leda clay in this part of Canada. On this terrace in many places the Saxicava sand forms the surface, and the Leda and boulder-clay may be seen beneath it. This may be called at Montreal the Sherbrooke Street terrace. Another, the Water-work terrace, is about 220 feet high, and is marked by an indentation on the lower Silurian limestone. At this level some boulder-clay appears, and in places the calcareous shales are decomposed to a great depth, evidencing long sub-aerial action. Three other terraces occur at heights of 386, 440, and 470 feet, and the latter has, at one place above the village of Côte des Neiges, a beach of sand

and gravel with Saxicava and other shells;* while, in a depression above this, at a height of 550 feet, sea-shells occur in clay and sand, and there is a distinct beach at about 615 to 625 feet.† Even on the top of the Mountain, at a height of about 700 feet, large travelled Laurentian boulders occur, lying loose and without any boulder-clay. On the lower St. Lawrence, below Quebec, the series of terraces is generally very distinctly marked, and for the most part the lower ones are cut into the boulder and Leda clays, which are here of great thickness. I give below rough measurements of the series as they occur at Les Eboulements, Little Mal bay and Murray bay, where they are very well displayed. I may remark in general, with respect to these terraces, that the physical conditions at the time when they were cut must have been much the same with those which exist at present, the appearances presented being very similar to those which would occur were the present beach to be elevated.

Comparisons of the older and modern terraces may be made at many places on the lower St. Lawrence. At Little Metis, where I have had good opportunities of studying their appearances, the coast is fringed with a broad belt of boulders, wholly covered at high tide, but exposed at low tide, and occupying in many places a breadth of thirty to fifty paces, within which the boulders are packed very closely. They vary in size from nine to ten feet in diameter downward, and consist principally of orthoclase gneiss, Labradorite rock and other crystalline rocks from the Laurentian of the north shore, here about thirty-five miles distant at the nearest point. With

* The beach at Côte des Neiges is that described by Sir C. Lyell in his travels in N. America.

† Dr. F. D. Adams and Baron de Geer.

these are masses of the hard sandstones of the lower
Silurian rock of the south coast, and occasionally, though
rarely, blocks of the upper Silurian limestone of the
inland hills to the south.

The boulders of this belt, though stationary in summer,
are often moved by the coast-ice in winter. This is well
seen where they have been removed to form tracks
for launching boats. In this case it is not unusual
to find in the spring that such tracks have been partially
refilled with boulders. On my own property, a track of
this kind was completely blocked a few years ago by an
angular boulder of sandstone nine feet in length, which
had been lifted from a spot a few feet distant; and it is
quite usual to find in a boat-track, cleared in the previous
summer, a dozen boulders of two feet or more in diameter
that have been dropped in it by the winter ice. Whether
any of these blocks are being drifted at the present time
from the north shore is not known; but they are moved
freely up and down the coast, and in dredging in depths
of eight to fifteen fathoms, I have found evidence that
large boulders are not uncommon on the bottom; and
judging from the small specimens taken up by the dredge,
they are similar to those on the shore, though apparently
with a larger proportion of flat, slaty fragments.

If the coast were now in process of subsidence, there
can be no question that the boulders would be pushed
upward, and would eventually form sheets and ridges of
boulders embedded in mud, much in the manner of the
marine boulder-clays now found inland.

Above high water, on certain portions of the coast,
there is a low terrace, only a few feet above the sea, and
consisting of sand, shingle, and gravel, often with fragments of marine shells. Boulders are not numerous on

this terrace, and are usually merely fragments from ledges of local sandstone. Bones of large whales occasionally occur on this terrace.

Proceeding inland, we find a second terrace about thirty feet above the sea, and consisting of sand, resting on hard boulder-clay or till. This last, at different places along the coast, is seen to vary in quality, being sometimes hard and loaded with boulders, in other cases a clay with marine shells, and again a clay with few boulders except at its junction with the sand above. On the inner side of this terrace, where it adjoins the rocky ledges inland, there is often a raised boulder-beach like that on the present shore, but with fewer and smaller boulders, as if the transporting power had been less than at present, and possibly the time of its action more limited. But still higher, on rocky ledges and gravel terraces, rising to the height of fifty to sixty feet, there are large Laurentian boulders, often forming inland boulder-beaches like that of the shore, and such inland beaches are found up to at least 400 feet. There are also a few upper Silurian boulders from the south, which become more numerous and larger further inland. In some places these Silurian limestone boulders are sufficiently numerous to afford the material for the supply of lime-kilns providing for local requirements. In some localities they would seem to be the deposits of glaciers descending from the hills to the south, but in others they would appear to have been water-borne.

The exposed ridges of rock on the second terrace and on the beach are sometimes polished with ice action, and show the normal N.E. and S.W. striation. I had no opportunity to observe the condition of the rock-surface under the boulder-clay. On the greater part of the

6

sixty-feet terrace, the rock-surfaces are rough, and yet large boulders often rest directly upon them.

The till or hard boulder-clay of this coast would be claimed by some glacialists as glacier work; but there can be no doubt that these clays locally contain marine shells, and there is therefore no need of invoking land-ice for their deposition. In this respect they agree with the drift-deposits of the lower St. Lawrence generally, except in the case of certain lateral valleys which seem to have been occupied with local glaciers descending from the Laurentian highlands.

TERRACES, NORTH SHORE OF LOWER ST. LAWRENCE.

Heights in English feet, roughly taken with Locke's level and aneroid.

LES EBOULEMENTS.	PETITE MAL BAY.	MURRAY BAY. W. Side.		E. Side.
900	——	892 .. ——	..	——
660	748	—— .. ——	..	——
479	505	—— .. 448	..	455
—	—	345 .. 378	..	346
325	318	—— .. 312	..	——
226	239	—— .. 281	..	259
—	—	136 .. 139	..	—
116	145	—— .. 116	..	127
—	——	50 .. 81	..	73
22	26	32 .. 30	..	—

Another series of levels taken by Mr. W. B. Dawson, along the road to Petit lac and beyond, gives the following heights:

	Feet.
Hill south of Petit lac, with drift and boulders at this level	1374
Drift ridge east of lake	810
Water level, Petit lac; appears to discharge over drift ridge or moraine	728
Clay, capped with 10 feet sand	589
Clay terrace	241
" " bank Murray Bay river	73

With reference to the differences in the above heights, it is to be observed that the terraces themselves slope somewhat, and are uneven, and that the principal terraces are sometimes complicated by minor ones dividing them into little steps. It is thus somewhat difficult to obtain accurate measurements. There seems, however, to be a general agreement of these terraces, and this I have no doubt will be found to prevail very extensively throughout the lower St. Lawrence. It will be seen that three of the principal terraces at Montreal correspond with three of those at Murray bay; and the following facts as to other parts of Canada, gleaned from the Reports of the Survey and from my own observations, will serve farther to illustrate this:

	Feet.
Kemptville, sand and littoral shells	250
Winchester, " " "	300
Kenyon, " " "	270
Lockiel, " " "	264 & 290
Hobbes' Falls, Fitzroy, sand and littoral shells	350
Durham Mills, De L'Isle, " " "	289
Upton	257

The evidence of sea action on many of these beaches, and the accumulation of shells on others, point to a somewhat long residence of the sea at several of the levels, and to the intermittent elevation of the land. On the wider terraces, at several levels it is usual to see a deposit of sand and gravel corresponding to the Saxicava sand. One of the most important terraces throughout the lower St. Lawrence is that between 500 and 600 feet, which seems to correspond with the time of deposition of the principal bed of fossiliferous Leda clay. Corresponding to the terraces on rising grounds are the "boars' backs," kaims or eskers stretching along flat lands between pro-

jecting hills, and following old lines of coast. These are
evidently of the nature of modern gravel and shingle
banks, and are distinguished from moraines and ice-shove
deposits by their water-worn and sorted material.

On the lower St. Lawrence I have observed marine
shells on the terraces up to about 600 feet above the level
of the sea, but they will probably be found by diligent
search at higher levels. In the arctic region, Captain
Fielden (Journal of Geol. Society of London, Vol. XXXIV.,
1878, p. 566) reports Pleistocene shells, viz., *Pecten
Islandicus, Astarte borealis, Mya truncata* and *Saxicava
rugosa*, at the height of 1,000 feet above the sea.

With the terraces and elevated banks must be associated
the later boulder-drift, which has distributed travelled
stones and boulders through and over the Saxicava sand
and the moraines of older local glaciers, and has deposited
them at high levels on hills and mountains far inland.
The assignment of such loose boulders to their precise
date is, however, often extremely difficult, a fact which
may be well seen from a study of the data accumulated by
the boulder committee of the Geological Society of Scotland, under the presidency of my friend, Mr. David Milne
Home. Neglecting altogether for the present boulders
not far removed from their native sites, some of the far-
travelled boulders at high levels may have been left as
residue of the denudation of the more elevated sheets or
patches of boulder-clay. Others may belong to the
driftage of the margins and banks of the mid-glacial
depression of the Leda clay, but these can scarcely have
reached higher levels than about 600 feet. Others still
may have been carried by ice in that short-lived depression
of very great magnitude which seems to have immediately

preceded the re-elevation of the Saxicava sand,* and it is
even possible that some may have been placed in their
present positions in the post-glacial subsidence, of which
there is evidence on both sides of the Atlantic. Some
belong to lake margins of post-glacial date. Thus no
general statement can safely be made respecting these
erratics, but each group or belt must be studied with
reference to its local associations and the source of the
material, as well as with reference to the probable stage
in the various continental subsidences and elevations to

Fig. 3.—Terraces at L'Anse à Loup, near Tadousac. Lower terrace, clay; upper, sand.

which it belongs. *The assumption that all boulder-drift
may belong to one period is a fertile source of error*, and
though many important observations on the subject have
been made by Spencer, Dr. G. M. Dawson, Chalmers and
others, there is an almost unlimited field for detailed work
in this direction.

A still farther complication arises here from the probability of differential elevation, whereby the relative
levels have been changed in different parts of the Pleisto-

---

* McGee refers to this in connection with the "Columbia" deposits
of the Appalachians.

cene, as illustrated by Dr. G. M. Dawson in his Memoirs
on British Columbia, by Mr. Chalmers on the lower St.
Lawrence, by Upham, Gilbert and Spencer in the lake
regions, and by De Geer in Sweden. To Spencer we
are indebted for a great mass of valuable observations on
the lake margins of the Canadian lakes and the questions
of the origin of the lakes and the primitive course of the
St. Lawrence river anterior to the Pleistocene age, as well
as to the former greater extension of the lakes and the
differential elevation by which they have been affected.[*]

---

[*] Canadian Naturalist, 1882. Trans. R.S. Canada, 1889. See also
Warren Upham's Appendix to Wright's Ice Age and McGee's Seventh
Report Am. Geol. Survey, p. 639.

CHAPTER III.

PHYSICAL AND CLIMATAL CONDITIONS.

I.—General Conditions.

It is, I think, universally admitted that the later Pliocene age, immediately preceding that of the boulder-clay, was a period of elevation of the continents in the northern hemisphere, the "*first* continental period" of Lyell. The evidences of this are to be found in every text-book of geology, and in Canada I may refer to the excavation of the Saguenay valley, as explained by me

Fig. 4.—Valley of Lower Saguenay—Old glacier bed.

in my notes of 1872, referred to below, and to the similar evidence accumulated by Dr. G. M. Dawson regarding the cañons of British Columbia.* It has also been conclu-

---

* See also Upham, Geol. Magazine, Nov., 1890, and Bul. Geol. Society Am., Vol. I. ; Spencer, *Ib.*, May, 1890 ; and Journal Geol. Society, Nov., 1890.

sively shown by several geologists * that in this period
the valleys of the great American lakes were excavated,
and that the ancient St. Lawrence flowed without any
lakes to the sea. The present great lakes are partly
dammed up by glacial deposits, and partly produced by
warping or differential elevation. It may now be considered as fully established that the great American lakes
are not the result of glacial action, but that they are old
river valleys excavated in periods of continental elevation,
and now dammed up by accumulations of *debris* and by
differential elevation occurring in the Pleistocene period.
In the great depression of that period, they spread
far more widely than at present, as indicated by the
old terraces around them, some of which, according to
Spencer, are 1,700 feet above the present water level, and
may indicate a period when the whole American land was
much lower than at present. (See Spencer, Journal
Geol. Society, Vol. XLVI., 1890.) Further, Dr. G. M.
Dawson † has shown that in this and previous periods of
continental elevation the great fiords and cañons of
British Columbia were cut out, and quite recently
Pettersen has ably applied the same explanation to the
fiords of Norway. The latter says: "I have, therefore,
after the most careful researches here, yard by yard, and
extending over many years, come to the conclusion *that
the Balsfjord is not of glacial origin, but formed an incision
or depression in the mountains of older origin than the
glacial age.* And this conclusion, I believe, *may, in the
main, apply to the question of the formation of all fjords in*

---

* Newberry, Hunt, Spencer.

† Superficial Geology of British Columbia, 1878. Later Physiographical Geology of the Rocky Mountains of Canada, Trans. R.S.C.,
1890.

the north of Norway.   But whether it is applicable to all fjords *in the whole of Norway* I shall not attempt to answer." *

I have, in my Pleistocene notes of 1872, taken the

Fig. 5.—Higher cliffs of Saguenay gorge.—Cape Eternity.

valley of the Saguenay as a typical illustration, and have shown that along an old fracture of the Laurentian rocks

---

* Nature, June, 1885.

fluvatile denudation in Pliocene and pre-Pliocene times
has cut a trench to a depth of 800 feet below the present
water level of the St. Lawrence, and that the glacial
action of the Pleistocene has polished and grooved its sides
and probably its bottom, and piled up *debris* at its mouth.*

I need hardly say, after the discussions on the subject,
that the reference of the cutting of lake basins and fiords
to glaciers in the ice age, against which I have argued
ever since 1866, has been altogether exploded.

(1) This being admitted, and also the fact established,
by the most convincing evidence, of the great depression
of our continents in the glacial or Pleistocene age, it
follows that the first or oldest of the Pleistocene deposits,
the till or boulder-clay, was laid down during a time of
subsidence, in which the northern land was slowly sinking
under the sea. We leave untouched at present the mode
of deposition of boulder-clay and of polishing and striation of rock-surfaces under it, merely noting that the
boulder-clay proper is confined to the plains and valleys,
where it often contains marine remains. The hills show
evidence of glacier-movement down their valleys, and
of the formation of moraines, and sometimes of patches
of an indurated ground moraine or hard till, different
from ordinary boulder-clay.

(2) The formation of the Leda clay and interglacial
deposits, and of the similar deposits on the western
plains, belongs to the time when this region had subsided
beneath the waters, with tracts and islands of higher
lands projecting. The differential character of this elevation, whereby certain parts of the then submerged areas
stood higher than others, will be mentioned later.

---

* Notes on Pleistocene of Canada, 1872.

At this period the valley of the St. Lawrence and the eastern coast, as far west as lake Champlain and the east end of lake Ontario, as well as the borders of the arctic basin and of the Pacific, were under the waters of the ocean and inhabited by a rich boreal fauna, nearly all the species of which, in its eastern development, I have myself dredged alive in the waters of the estuary of the St. Lawrence. On the other hand, the western plains were covered with waters which have not afforded marine animals in their deposits, but hold remains of land plants. Farther, these land plants were of species not arctic, but merely boreal or north temperate,* while the proper arctic flora must have been still farther north.

(3) This mid-glacial period was followed by the second boulder-deposit, in which still farther subsidence occurred, and boulders were carried by floating ice to the summit of the higher hills in eastern Canada and New England, up to the height of 4,200 feet, and in the Rocky Mountains even to the great elevation of 5,289 feet.† This second period of boulder-drift and its deposits must not be confounded with the earlier till.

(4) From this depression the continent arose gradually or by intermittent throes, leaving the terraces of the hills and the sand and gravel beds (Saxicava sand) of the plains as evidences of the recession of the waters. This elevation proceeded so far as to inaugurate the *second* continental period, when the land was more extensive than at present, and a southern fauna penetrated far north along our coasts, while great mammals, now extinct, overspread the land. Since that time there have been

---

* See Chapter V.
† G. M. Dawson, Report on Superficial Deposit, Bow river, 1884.

cataclymic oscillations of level and a partial subsidence, which is apparently still in progress.*

For the evidence of this history I may refer to the papers cited in the notes, † where abundant facts will be found relating more especially to Canada, and, so far as my reading extends, they will be found applicable, with certain modifications of details, to other parts of the northern hemisphere.

In closing this section, I desire to refer to the map (Fig. 6, B.) of the geography of North America in the early Pleistocene, the height of the glacial period. At this time I believe the northern half of North America consisted of three large and mountainous islands, clad for the most part with nevé and glaciers, and surrounded by ice-laden seas and straits. The conditions, it will be seen at a glance, were most favourable to refrigeration, by accumulation of floating ice in temperate latitudes, while the arctic climate may have been little more severe than at present, and the extreme opposite of those which existed in the warm period of the early tertiary, when the northern end of the continent was closed against the arctic currents, and when the interior continental plateau constituted a northern extension of the warm waters of the gulf of Mexico. This map implies differential depression of the western plains as compared with the mountains, and of the northern as compared with the southern portions of North America, and an opening for

---

* According to Merrill and Lendenkehl (American Journal Science, June, 1891), alternate depression to the amount of 150 feet and elevation to the amount of 400 feet have occurred in the valley of the Hudson river since the glacial period. See also Acadian Geology, article "Submarine Forests."

† Also notes on Canadian Pleistocene, 1872; Acadian Geology, 1878.

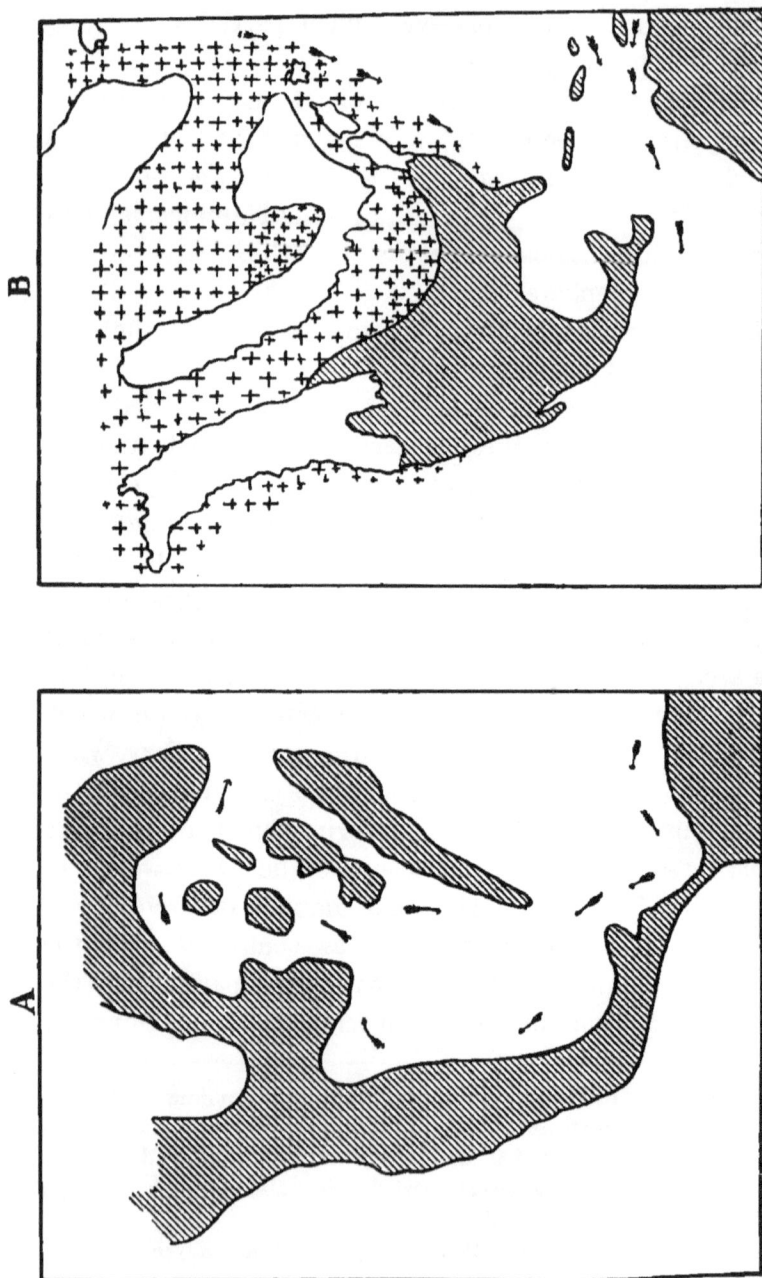

Fig. 6.—America in the Cretaceous (A) and Pleistocene (B).
Shaded portions, land; crosses, ice-laden sea; unshaded bands, glaciated mountains.

the equatorial current between North and South America, all of which suppositions are substantiated by known geological facts, more especially the occurrence of Pleistocene fossils at high levels, and of the same species of modern shells on the Atlantic and Pacific shores. Following the example of those geologists of the United States who are in the habit of giving a factitious reality to their palæogeographical views, by attaching names to extinct lakes, etc., we may name some of the more prominent features of our map after eminent living advocates of extreme glacial views, whose personal merit and ability, I am prepared to admit, are in the inverse proportion to the probability of their theoretical views. The great southern bay, at the bottom of which lies the "terminal moraine," may bear the name of Dana. The strait leading to the north-east, where the St. Lawrence now flows, may be Upham strait. The great western opening may well be called Chamberlain sound, and the northern bay, filled with ice in the region now occupied by Hudson's bay, may be the gulf of Wright. The greater islands will be respectively Cordilleran and Laurentide lands, fit companions of Greenland; and the smaller eastern island, Appalachia Infelix. Thus will be completed the rough general outline of one map of America in the age of the boulder-clay. Respecting Dana bay and Wright bay on the map, it is evident that the heavy ice-fields borne down by the arctic current and north-west gales, and the bergs derived from the mountain glaciers, would choke them with continuous masses of ice of enormous width, the pressure of which would pile up heaps of broken ice full of stones and earth on their shores, and would exert a mechanical force much greater than that of ordinary glaciers, so that morainic accumulations of great magnitude would be produced of the same general nature with those of the Missouri coteau. These, of course, now constitute the great "terminal moraine" * which has been so carefully traced by the geologists of the United States.

Comparing a map of America in the Pleistocene with that of the same region in the later Cretaceous and early Eocene, it will be at once seen how, in the one case, the arctic conditions must have been transferred to temperate regions, and how, in the other, temperate conditions must have been carried north to Greenland.

It may be well here to notice shortly the contention often made that the weight of the ice upon the parts of the continents loaded with it must have been itself a cause of the Pleistocene depression. No one has, I believe, contended more strenuously than the writer, in connection both with the carboniferous deposits and those of the deltas of great rivers,† in favour of the instability of the crust of the earth when loaded with great weights or when these are removed; but it must be observed that such weights are usually due to the deposition of sediment in the sea. The effect of accumulations of ice on high lands is less certainly known. If, however, we imagine that the continental period of the later Pliocene was closed by a differential depression, submerging the plains and leaving the mountains elevated, the resulting geographical conditions would be favourable to accumulation of

---

* I need scarcely say that the reference of this terminal moraine to a land glacier is absurd on physical grounds; and there is no modern example of such a thing, as even the Greenland névé discharges by local glaciers.

† "Acadian Geology," "Modern Science in Bible Lands," Presidential Address to Brit. Association, 1886.

snow and ice on the latter. But if the pressure of such snow and ice was sufficient to depress the hills, it must necessarily at the same time elevate the plains, and this change, by diminishing evaporation and by increasing continental warmth, would at once cause the ice-caps to melt away. Thus subsidence produced by accumulations of ice would at once accomplish the destruction of such accumulations, while it would remove the high lands necessary to account for any extensive movement of glacial ice. In other words, as elsewhere urged in this volume, the facts of dynamical geology and physical geography are fatal to hypotheses of polar ice-caps and continental ice-sheets, and if one were to admit all that has been alleged in reference to depression of land by pressure of ice, these difficulties would not be removed in the slightest degree.

II.—*Causes of Glaciation and Distribution of Erratics.*

We now come to consider the probable causes of glaciation and boulder-deposits, and first the agency of "continental" and local glaciers.

1.—GLACIER ACTION, ETC.

1. With respect to the agency of land ice, I have no hesitation in saying that, as I have maintained for thirty years, a sheet of ice covering the wide surface of the American continent, and piling up a "moraine" such as that which extends from the northern end of the Missouri coteau and south of the great lakes to the Atlantic coast, is a physical impossibility. It is so, first, because the only possible gathering-ground of sufficient snow to form glaciers is on high lands sufficiently cold and sufficiently

near to the ocean to receive and condense its burden of
watery vapour.   This is the cause of the present state of
Greenland, and similar conditions would account for that
great Cordilleran glacier which Dr. G. M. Dawson has
shown existed in Pleistocene times on the mountains of
British Columbia, and for the Laurentide glacier or local
glaciers which it is known existed on the Laurentian [*]
highlands of Canada and even on the extension of the
Appalachian mountains in eastern Canada.[†]   On this
subject I may quote here the conclusions of the well-
known Russian geographer, Von Woeickoff,[‡] as summarized
in a partial translation published in the "Canadian
Naturalist" in 1882.   I ought perhaps to apologize for
repeating here common and even trite conclusions of
physical geography; but my excuse must be the neglect
with which they have been treated by so many geologists,
and the extent to which theories altogether at variance
with them have been promulgated.

I may say at the outset that I fully agree with the
views as to the *motion* of glaciers contained in the sub-
joined extract :

"The fuller consideration of the physical properties of
glacier ice leads to essentially the same conclusions as
those to which Forbes was led forty-one years ago, by the
study of the larger phenomena of glacier motion, that is,
that the motion is that of a slightly viscous mass, partly
sliding upon its bed, partly shearing upon itself under the

---

[*] Notes on Post-pliocene, 1872.   See also a paper by McGee in the
the Proc. American Association (Boston, 1880, p. 447), and Dana in
American Journal of Science, 1872.

[†] Chalmers' Glaciation of N. New Brunswick, etc., Trans. R.S.C.
1886.

[‡] Geological Society, Berlin, 1881.

7

influence of gravity." (Trotter, in Proc. Royal Society of London, XXXVIII., 107.)

Woeickoff's conclusions may be summed up as follows:

" 1. The great expanse of ocean in the southern hemisphere is favourable to the deposit of snow and formation of glaciers, by furnishing a great evaporating surface, and at the same time a low general temperature facilitating precipitation. This applies to the antarctic continent, and also permits the formation of glaciers far to the north in New Zealand and in South America.

" 2. On the other hand, the present condition of the northern hemisphere is unfavourable to glaciers, because the sea is so warm that deposition near the coasts is rather as rain than snow up to pretty high latitudes, while the continents are so wide that there is little precipitation in their interior.

" 3. Thus there are no glaciers in eastern Siberia, even in the mountains, where the mean temperature is only 15ᶜ to 16° C., and central Asia generally is unfavourable to glaciation on account of its dryness, while eastern Asia is acted on by the monsoons. If, therefore, the extent of land in Asia has not materially changed since the Pliocene period, there could not have been great glaciers there since that period. Even the submergence of the great plain of China could not materially affect this result, though it might cause glaciers in the mountains of Japan.

" 4. To explain the great Pleistocene glaciers, of which traces are found in western Europe, it is necessary to suppose that the temperature was lower, either on account of submergence of the low lands or of diversion of warm currents, or both causes may have operated. A submergence connecting the White and Baltic seas would greatly promote the production of snow and ice.   But this could not affect the interior of Russia or of Asia, so long as their plains remained above water.

" 5. The submergence of the plains must be a necessary condition of the general glaciation of the higher lands.

" 6. Astonomical changes do not affect this result. With a great eccentricity of the orbit and the winter in aphelion the colder winters and hotter summers would produce more powerful monsoons, while on the opposite condition the interior of the continents would have warmer winters and cooler summers and weaker monsoons. In either case the conditions for continental glaciers would not be improved.

" 7. These considerations show that general coverings of ice stretching from the Pole to perhaps 45° are impossible. Under conditions of submergence of the plains the sea must keep open, in order to afford material for snow on the remaining high lands, and with large continental plains the climate will be too dry for glaciers.   Thus there must always be seas free from ice, or continental plains free of ice, and under most supposable conditions there must be both."

The following comments by the writer accompanied the above abstract in 1882 :

Applying these very simple geographical truths to the North American continent, it is easy to perceive that no amount of refrigeration could produce a continental glacier, because there could not be sufficient evaporation and precipitation to afford the necessary snow in the interior.   The case of Greenland is often referred to, but this is the case of a high mass of cold land with sea mostly open on both sides of it, giving, therefore, the conditions most favourable to precipitation of snow.   If

Greenland were less elevated, or if there were dry plains around it, the case would be quite different; as Nares has well shown in the case of Grinnel land, which in the immediate vicinity of Greenland presents very different conditions as to glaciation and climate.

If the plains were submerged and the arctic current allowed free access to the interior of the continent of America, it is conceivable that the mountainous regions remaining out of the water should be covered with snow and ice, and there is the best evidence that this actually occurred in the glacial period; but with the plains out of water, there could never have been a sufficiency of snow to cause any general glaciation of the interior. We see evidence of this at the present day in the fact that in unusually cold winters the great precipitation of snow takes place south of Canada, leaving the north comparatively bare, while as the temperature becomes milder the area of snow deposit moves further to the north.

The writer has always maintained these conclusions on general geographical grounds, as well as on the evidence afforded by the Pleistocene deposits of Canada, and he continues to regard the supposed evidence of a terminal moraine of the great continental glacier as nothing but the southern limit of the ice-drift of a period of submergence. In such a period the southern margin of an ice-laden sea where its floe-ice and bergs grounded, or where its ice was rapidly melted by warmer water, and where consequently its burden of boulders and other *débris* was deposited, would necessarily present the aspect of a moraine, which by the long continuance of such conditions might assume gigantic dimensions. Some anomalies in the levels of the so-called terminal moraine are no doubt due to differential elevation.

By many writers on this subject it is apparently maintained that in North America a continental glacier extended in temperate latitudes from sea to sea, and this glacier must, in many places at least, have exceeded a mile in thickness. Independently of the physical difficulties attending the movement of such a mass without any adequate slope, it is obvious, from the considerations above stated, that the amount of snow necessary to the production of such a glacier could not possibly be obtained. With a depression such as we know to have existed, admitting the arctic currents along the St. Lawrence valley, through gaps in the Laurentian watershed, and down the great plains between the Laurentian areas and the Rocky mountains, we can easily understand the covering of the hills of eastern Canada and New England with ice and snow, and a similar covering of the mountains of the west coast; more especially when we take into account the probability of an elevation of the mountains along with the depression of the plains, and of the southern part of the continent not having been depressed, and so blocking the exit of the ice to the south, along with the escape of the equatorial current through the isthmus of Panama, then submerged. The sea also in this case might be ice-laden and boulder-bearing as far south as 40°, while there might still be low islands far to the north, on which vegetation and animals continued to exist. We should thus have the conditions necessary to explain all the anomalies of the glacial deposits.

Whatever difficulties may attend such a supposition, they are small compared with those attendant on the belief of a continental glacier, moving without the aid of gravity, and depending for its material on the precipitation taking place on the interior plains of a great continent.

On the other hand, the evidence of great local glaciers in the Pleistocene period is of the best possible description. I may refer here to the indications obtained by Dr. G. M. Dawson of an immense glacier or group of glaciers occupying the Cordillera of British Columbia, and discharging its ice to the north into the Yukon valley, to the south into Puget sound, and to the west into the Pacific. Here, as he has shown, the conditions were combined of a high mountain chain with the Pacific on the west, and the then submerged area of the great plains on the east, affording next to Greenland the grandest gathering-ground for snow and ice that the northern hemisphere has seen.

The movement of ice north and south from the old gathering-ground of the Laurentian axis has been shown by the reports of arctic explorers and of the geological survey.* That from the Notre Dame mountains on the south side of the St. Lawrence, as shown by Chalmers,† and the radiation of ice from the central districts of Newfoundland, as described by Kerr and Murray,‡ are other examples.

Thus the existence of local glaciers on the west and east and on the higher lands facing the north has been established, and this not merely in the later, but in the earlier Pleistocene; but whatever of increase or diminution they experienced in the course of that period, they could never have become a continental glacier, spreading over the plains, nor could there have been a polar ice-cap,

---

* See papers and reports already referred to, by Dawson, Bell and others.

† Trans. R.S.C., 1886.

‡ Journal Geol. Society, 1876. Reports and Survey of Newfoundland.

since the facts obtained by the Canadian survey and the arctic explorers show that these local glaciers discharged ice and bergs both to the north and south. Some of the evidence of this is thus stated by Dr. G. M. Dawson.*

"Along the arctic coast, and among the islands of the archipelago, there is a considerable volume of evidence to show that the main direction of movement of erratics was *northward*. Thus, boulders of granite, supposed by Prof. Haughton to be derived from North Somerset, are found 100 miles to the north-eastward (Appendix to M'Clintock's Voyage, p. 374), and pebbles of granite, identical with that of Granite point, also in North Somerset, occur 135 knots to the north-west (*op. cit.*, p. 476). The east side of King William Land is also said to be strewn with boulders like the gneiss of Montreal island, to the southward (*op. cit.*, p. 377). Prof. Haughton indicates the direction and distance of travel of some of these fragments by arrows on his geological map of the arctic archipelago, and reverts to the same subject on pages 393, 394, pointing out the general northward movement of ice indicated, and referring the carriage of the boulders to floating ice of the glacial period.

" Near Princess Royal island, in Prince of Wales strait, and also on the coast of Prince of Wales island, the copper said to be picked up in large masses by the Esquimaux (DeRance, Nature, Vol. XI., p. 492), may be supposed to be derived from the Cambrian rocks of the Coppermine river region, to the south, as it is scarcely possible that it occurs in place anywhere in the region of horizontal limestone where it is found.

---

* Notes on the geology of the northern part of Canada. Geol. Survey of Canada, 1886.

"Dr. Armstrong, previously quoted, notes the occurrence of granitic and other crystalline rocks, not only on the south shore of Baring island, but also on the hills inland. These, from what is now known of the region, can scarcely be supposed to have come from elsewhere than the continental land to the southward.

"In an account of the scientific results of the 'Polaris' expedition (Nature, Vol. IX.), it is stated of the west coast of Smith's sound, north of the Humboldt glacier, that 'wherever the locality was favourable, the land is covered by drift, sometimes containing very characteristic lithological specimens, the identification of which with rocks of South Greenland was a very readily accomplished task. For instance, garnets of unusually large size were found in latitude 81° 3', having marked minerological characters by which the identity with some garnets from Tiskernaes was established. Drawing a conclusion from such observations, it became evident that the main line of the drift, indicating the direction of its motion, runs from south to north.' It should be stated, however, that Dr. Bessels, who accompanied the 'Polaris' expedition, regards these erratics as certainly not transported by glaciers, but by floating ice, and as showing that the current of Davis' strait was formerly to the north, and not to the south, as at present. (Quoted from *Bulletin de la Soc. de Géographie*, Paris, March, 1885, in Arctic Manual, p. 553.)

"It may be mentioned, as bearing on the general question here referred to, that Dr. Bell has found evidence of a northward or north-eastward movement of glacier ice in the northern part of Hudson bay (Annual Report Geol. Survey of Canada, 1885, p. 14, DD), with distinct indications of eastward glaciation throughout Hudson strait. (Report of Progress, Geol. Survey of Canada, 1882–84, p. 36, DD.)

" The facts available for this northern part of the continent and the Arctic islands thus rather point to a movement of ice outward in all directions from the great Laurentian axis or plateau, which extends from Labrador round the southern extremity of Hudson bay to the Arctic sea, than to any general flow of ice from north to south, from the vicinity of the geographical pole."

The same writer, in his more recent paper on the Rocky mountains, refers to the facts, that while the Cordilleran glacier discharged to the northward, the McKenzie River valley is shown to present similar phenomena, and that the absence of drift in the northern part of the Yukon district and along the arctic coast as far as the McKenzie river (Dease and Simpson) shows that this region may have been land enjoying a moderate climate at the same early Pleistocene period in which the mountains of British Columbia were covered with ice.*

The observations of Dr. G. M. Dawson in the Cordilleran region of British Columbia, already mentioned, are so important in this regard,† and are presented in so compact and clear a form, that I may be excused for quoting from his account of the great Cordilleran glacier of the west, which may be regarded as a specimen of those great local glaciers which accumulated in Pleistocene times on all the high mountains near the coasts of the continents, or which were surrounded by submerged plains capable of affording vapours to be precipitated upon their summits.   Let it be observed in this connection that the plains east of the Rocky mountains were

---

* See also observations of Mr. R. G. McConnell, Bul. Geol. Soc. America, Vol. I., and I. C. Russell, in the same volume.

† Later Physiographical Geology of the Rocky Mountain Region in Canada, Trans. R.S.C., 1890.

submerged at this time, otherwise the Cordilleran glacier
could scarcely have existed.

"The Cordilleran region, in consequence of its high
elevation, and probably also in part from other concurrent
causes by which the northern hemisphere was affected at
the inception of the period of glaciation, appears to have
become at this time pre-eminently the condenser of the
North Pacific. Precipitation occurred upon it chiefly in
the form of snow, which was so much in excess of the
influence of the summer heat as to accumulate from year
to year. Great glaciers formed in the higher mountains,
probably in the first instance among those situated nearest
to the coast; but eventually the greater part of the
region became covered and buried either in *névé* or
beneath glacier-ice. The directions of motion of the
glaciers at first produced was doubtless in conformity
with that of the valleys of mountain streams; but at a
later date, when the Cordillera became completely buried,
a general movement was initiated from a region situated
between the fifty-fifth and fifty-ninth parallels of north
latitude, in south-easterly and north-westerly bearings.*
The Cordillera, in fact, between the forty-eighth and
sixty-third parallels—or for a length of about 1,200
miles—seems to have assumed an appearance closely
analagous to that of Greenland at the present day, save
that in consequence of the high bordering mountain
ranges, with the general trend of these and of the lower
intervening country of the interior plateau, the greater
part of the ice was forced in this case to follow its length
in the directions above indicated, instead of discharging

---

* Such general movement probably affected only the central portion
of the ice-mass by which the Cordillera was covered, and there is no
reason to suppose that it was otherwise than sluggish.

laterally on both sides to the sea. A certain proportion of the ice, however, during the maximum phase of this great glacier, flowed through passes in the coast ranges, and uniting there with ice derived from the western slopes of these ranges, filled the great valley between Vancouver island and the mainland, impinged upon the shores of the Queen Charlotte islands, and still further north reached the ocean across the coast archipelago of the south-eastern coast-strip of Alaska.

"Having, from an examination of the notes of various arctic explorers, arrived definitely at the conclusion that the glaciers of the eastern part of the continent possessed a northward as well as a southward direction of motion from their main gathering-ground,[*] the writer was pleased to be able to avail himself of the opportunity afforded by the Yukon expedition to investigate the conditions of the northern part of the Cordilleran glacier.[†] Evidence was there obtained of its northward or north-westward direction of movement, and this has since been confirmed and added to by observations in surrounding regions by Mr. R. G. McConnell, of the Canadian Geological Survey (1888), and by Mr. I. C. Russell, of the United States Geological Survey (1889).[‡] On the Lewis and Pelly rivers, branches of the great Yukon, striated rock-surfaces, evidently due to the general Cordilleran glacier, were noted; in the case of the first-mentioned river as

---

    * Annual Report Geol. Surv. Can., 1886, p. 56, R.

    † Chalmers (Am. Geologist, Nov., 1890) very properly proposes the names Cordilleran, Laurentian, &c., "*System of Glaciers*," to express the fact that like the modern glaciers of the Alps and even of Greenland, a system and not a single glacier is meant.

    ‡ Bulletin Geol. Soc. Am., Vol. I., p. 99.

    § Geological Magazine, Dec. 3, Vol. V., p. 348. Annual Report Geol. Surv. Can., 1887-88, p. 40, B.

far north as latitude 61° 40', on the Pelly to latitude
62° 30', longitude 135° 45'. The observed bearings show
a convergence of direction toward the low country about
the confluence of these two rivers, near the site of old fort
Selkirk, and it is not improbable that the glaciers may
have here reached to the vicinity of the sixty-third
parallel on the one hundred and thirty-seventh meridian.
No traces of glaciation were observed by Mr. McConnell,
still further north, along the Porcupine river, nor by Mr.
Russell further down the main valley of the Yukon,* the
appearances there being, on the contrary, those of a
country which had long been subjected to subaërial decay,
and which had not been passed over either by glaciers
or by floating ice capable of bearing erratics.

"Further illustration of the fact that the extreme
north-western part of the continent remained a land-
surface upon which no extensive glaciers were developed
even during the time of maximum glaciation, is afforded
by the note of Messrs. Dease and Simpson as to the
entire absence of boulders along the arctic coast westward
from the estuary of the Mackenzie river.†

"Granting that the north-western extremity of the
Cordilleran glacier reached the furthest point above
assigned to it, we find that its extension, from the central
gathering-ground (or from the approximate margin of
this gathering-ground already given), was much shorter
than that obtained by the south-easterly flowing part, the
approximate lengths being 350 and 600 miles respectively.
This may be regarded as indicating either a greater
relative elevation of this part of the continent to the

---

* Bulletin Geol. Soc. Am., p. 140.
† Narrative of Discoveries on the North Coast of America, 1836-39,
p. 149.

north-westward, or a less copious supply of snow in that direction, the latter being the more probable supposition on account of the absence, which has just been referred to, of traces of glaciation in the extreme north-west.

"The flooding of the great plains by arctic waters while the Cordillera stood as a much-elevated land between them and the warmer waters of the Pacific, would in itself go far to explain the conditions under which the excessive precipitation required for the production of the great Cordilleran glacier occurred."

It would appear that there is no evidence of any extensive flow of ice from the Cordilleran glacier to the eastward, though small local glaciers may have moved down to the submerged plains, which, owing to the prevalent westerly winds, could contribute little to the snow on the mountain ranges. It would also appear that the maximum condensation of the Pacific moisture occurred not more than 200 miles inland from the west coast. At the same time, the submergence of the great plains of the St. Lawrence valley and of the eastern coast would place the Laurentide and Appalachian mountains under similar conditions, but not of so extreme magnitude.

We have thus a perfect geographical picture of extensive local glaciation in the early glacial age, in circumstances most favourable for its existence. Let us not forget, however, that this picture belongs only to one portion of the period. There was in the Cordillera also a mid-glacial period of subsidence of the mountain axis, and a later development of smaller local glaciers, the deposits of which have also been worked out by Dr. G. M. Dawson.

Two very interesting and important series of observations bearing on the glaciation of the eastern part of

Canada appear in the Report of the Geological Survey
published in 1879.*

Dr. Ells, in his Report on the Eastern Townships of
the province of Quebec, notes the fact that the glaciation
along the western side of the hills of that region, constituting an extension of the Appalachian chain, follows
the valleys, the striation pointing *westward toward the St.
Lawrence valley*. This is the old glaciation appertaining
to the till or boulder-clay, and proves local as opposed to
continental glaciation. On the other hand, he states that
numerous boulders of Laurentian rock, carried in the
opposite direction across the wide and deep St. Lawrence
valley, lie on the sides of the hills up to an elevation of
1,500 feet. Sea shells are also found, though only at a
few localities and not so high. This high-level drift
belongs to the second glacial period. Dr. Ells recognizes
this, and very properly refers the lower or till deposit to
local land glaciers, the newer and higher drift to submergence and floating ice. He also refers to the fact
stated in Logan's Geology of Canada, that raised beaches
occur at Ripton in the Green mountains at an elevation
of 2,196 feet.†

With these observations of Ells may be placed those of
Upham and other American geologists, and more recently
of Shaler in Maine, on the other side of the mountain
range. In his report on mount Desert, he shows that the
movement of ice was, as elsewhere on the New England
coast, to the south-eastward, or from the Appalachian
range—the converse direction to that found by Ells on
the opposite side. Shaler also finds the underlying till,

---

* Report for 1887-8.

† These observations perfectly agree with those of Chalmers and
the writer, already mentioned.

the intermediate Leda clay, and the overlying sea-drift, the latter implying, if rightly interpreted by him, a subsidence of at least 1,300 feet.* McGee, in the reports of the same survey,† in describing the so-called " Columbian " formation farther south, details facts of similar import; while in Chamberlain's map, in the same report, the direction of striation leads to the same conclusion, except that on the west side of the Appalachians *the arrows are reversed*, making the glaciers move up the mountain instead of down. Other citations might be made to show that the geologists of the United States, while still adhering to the hypothesis of a continental glacier with a terminal moraine, stretching half-way across North America, are accumulating facts in accordance with the results worked out by Canadian geologists in the Northern Appalachians, in the Laurentides and the Western Cordillera, and which must eventually greatly modify their views.

In the same report of the Canadian Survey with the observations of Dr. Ells, those of Mr. Low, in his report on James's bay, show that on the east side of the bay the glacial striae indicate a movement of ice to the westward from the high Laurentian land east of the bay. Thus the great V-shaped Laurentian axis, while throwing off ice to the St. Lawrence valley on the south-east, and to the great plains then submerged on the south-west, was also pouring off its ice into the interior basin of Hudson's bay and into the arctic sea. It may, at the period of extreme glaciation, have filled Hudson's bay with ice, and there is evidence of a terminal moraine along the middle of the bay, which may have belonged to the interglacial

---

* Report U.S. Geol. Survey, Vol. VIII.
† *Ib.*, 1885-6.

period. Mr. Low also finds that the newer stratified gravels belonging to the later period of ice-drift extend to heights of probably 700 feet, indicating the great depression which succeeded the earlier glacial period.

All these observations, combined with others detailed in different parts of this work, constitute the most complete proof that in Canada the condition of the continent in the more extreme glacial period was one of local glaciers, principally on the greater mountain regions, with submerged plains and sea-drift intervening, and that this was succeeded by a time of partial elevation of the plains and diminution of the height of the mountains, followed by a great and very general submergence, with much ice-drift preceding the second continental or post-glacial period.

Dr. Otto Torell, Director of the Geological Survey of Sweden, is one of the ablest Scandinavian students of Pleistocene deposits, and has always been an advocate for the theory of land glaciation, in so far as north-western Europe is concerned. He has also had the advantage of visiting portions of North America. I had the pleasure several years ago of guiding him to the best exposures of glacial and drift-deposits near Montreal. After his return from America, he stated his general conclusions respecting our Pleistocene deposits in a short paper published in the Proceedings of the Royal Academy of Sweden.* In this he clearly states the conditions necessary to the production of glaciers as follows :

" 1. Abundance of moisture in the atmosphere.

" 2. A low temperature, due either to great elevations in low latitudes, or to high latitudes with or without such elevations of land.

---

* April, 1877.

" These conditions insure such accumulations of snow above the line of perpetual frost as will sooner or later descend below the line of perpetual snow and .be changed to ice and water.

" The water forms glacial rivers, and the ice will move as a plastic mass to a line determined by the amount of snow on the one hand and the climate on the other.

" The advancing movement of the glacier is accompanied by erosion and scratching of the rocks below and by the formation of different kinds of moraines, as *till* or blue boulder-clay, and yellow unstratified masses—terminal, lateral and superficial moraines. Simultaneous with these phenomena, we have the action of the glacial rivers, consisting in a partial denudation of the moraines, and the formation of stratified gravel, sand and clay."

He next explains his own views of the glaciation and dispersion of erratics from Scandinavia as a centre by the movement of glaciers, and applies these to America, admitting, however, that here there must have been separate centres of dispersion in the east and west. He thus states the objections to the current views of American land-glacialists :

" It has been the opinion of many distinguished American geologists that the source of the eastern ice-field is to be searched for in the Canadian highland. Against this opinion several important reasons may be urged. First, in those parts of Canada in which the glaciers in question are supposed to have originated, we have reason to believe that the rocks are *rounded* and *scratched*, phenomena everywhere recognized as glacial, but, I think, in no case characterizing rocks known to have been covered with perpetual snow.

8

"Again, the elevation and extent of the highest portions of Canada are hardly sufficient to account for the requisite accumulation of snow and ice. And, finally, so far as I have learned, there is not found upon the rocks of the northern slope of Canada, nor in boulders moved by glacial force, any satisfactory evidence that there has been a northward as well as southward movement of glaciers from the highlands of Canada." *

Refusing, however, to take into account the Pleistocene depression and the agency of floating ice, he finds himself under the necessity of adopting Greenland as the focus of dispersion and movement of glaciers for north-eastern America.

"If, therefore, the phenomena of the northern and eastern United States usually supposed to be glacial are indeed such, and if there is not sufficient reason for assuming the Canadian highlands to have been the source of the glaciers which produced these phenomena, then their source must be found elsewhere. I think it will be conceded by all geologists who have studied the glacial phenomena of these regions, that both the character of the erratics and the direction of the scratches upon the rocks show that this source must lie to the north-east. Following the line of glacial movement across Baffin's bay and Davis' strait into Greenland, we find the largest body of land in the northern hemisphere covered by ice and snow to a depth of not less than 2,000 feet, and at this moment sending down its icebergs as far as the Middle Atlantic.

"From the sixtieth degree of latitude to above the eightieth, this vast area of land is known to be ice-

---

* As already stated, later observations furnish this evidence.

covered, and from the scarcity of the icebergs on the eastern compared with the western coast of that land, it may be concluded that the general slope of the surface is to the south-west, and in the exact direction of the glacial markings and of what is known to have been the course of transported boulders in north-eastern America.

" Moreover, if we bear in mind the ascertained fact that during the glacial period the glaciers moving from the heights of Greenland toward the sea could not have formed detached icebergs, as now, but must have for the time blocked up all avenues except the one of easiest escape for the immense accumulation of ice, we may reasonably assume that this avenue was south-westward directly across British America and the north-eastern parts of the United States."

Probably few even of the more extreme glacialists will think this explanation at all feasible, and Torell himself, if acquainted with the additional facts ascertained in recent years, would probably see that it is unnecessary to go so far for the sources of ice. This will appear clearly when we consider the following leading facts already referred to in preceding pages :

1. There is no such universal glaciation of higher summits as that supposed. On these summits glaciated surfaces are rare and not strongly marked, and are most distinctly seen in the valleys and plains, to which also the boulder formation and till are for the most part limited.

2. There are the best reasons to believe that in the Pleistocene period the Laurentian highlands were proportionally more elevated than at present.

3. We now know with certainty that the prevalent drift on the north side of the Laurentian axis, as well as

in that of the great Cordilleran range, was to the north-ward, which removes Torell's most important objection.

4. The great facts of subsidence to a large extent in the Pleistocene period, the effects of this on climate and the certainty of the most extensive possible action of floating ice, are not referred to in Torell's theory, and would of themselves remove his objections, more especially when we consider the probability that depression of the plains was accompanied by elevation of the mountains.

It can scarcely be doubted that had these considerations been before the mind of the Swedish geologist, along with the admitted limitation of glaciers to centres in the east and west, his general conclusions would coincide very nearly with those stated in the previous pages with reference to the Cordilleran and Laurentian systems of glaciers and ice-drift over the lower levels.

It seems recently to have been ascertained that, in Norway, the old beach-lines are highest in the middle of the peninsula of Scandinavia, and descend toward the sea level at the extremities. This would seem to indicate that there, as in America, the older sea terraces belong to the time of the greatest glaciation, and that, if the ice weighed down the land in the manner indicated by these terraces, there could be no height of land maintained to send land-ice over the continent of Europe so far as claimed by Scandinavian glacialists.*

The results of the investigations of the "Challenger" in the antarctic ocean are of great importance with reference to the formation of marine till and stony clays. The dredge may now indeed be said to have settled this

---

* Matthew has referred to this in connection with the Pleistocene of New Brunswick in a paper in the Canadian Naturalist, VIII., p. 116, 1878.

question by ascertaining the deposition of marine boulder-clay, or at least of a deposit of sand and clay, with fragments of various rocks over areas perhaps as great as those now covered with similar deposits in the northern hemisphere. It is most instructive to find that a bed of this stony mud is in process of deposition from floating ice in the southern ocean, and this with such rapidity, that the foraminifera and other organisms elsewhere forming the deep-sea ooze are quite masked by it, while it is also possible that in some places all traces of these may be dissolved out by carbonic acid. It is further interesting to find such deposition taking place so extensively under conditions probably much less favourable than those which prevailed in Europe and America during the great Pleistocene subsidence.

These facts fully confirm the conclusions stated above with reference to the boulder-clay or till of North America, and which I have endeavoured to establish by the nature of the deposits now forming in the areas of ice-drift of the American coast, by the distribution and chemical condition of the boulder-clay itself, and by the occurrence of marine organisms in it. It is to be hoped that in future we shall not have so confident assertion that these remarkable clays are due to the direct action of land ice on the surface of our continents.

If the bottom of the South Pacific and Antarctic oceans could be elevated into land, we should see the evidence of glacier action on the hills representing the islands now out of water and extending from these a vast area of boulder-clay reaching as far north as our similar records of the Pleistocene submergence spread to the south, and probably holding in many places marine organic remains, though there might be expected to be few, both on account of the conditions of deposit and the solvent action of carbon dioxide in the cold bottom waters of the ocean. The only point wanting to complete the analogy would be that embaying and detention of drift-ice shown on our Pleistocene map, and by which the old glacial age was aggravated both in Europe and North America. To produce this would require some differential elevation of the bed of the Pacific toward the northern margin of the present area of ice-drift.

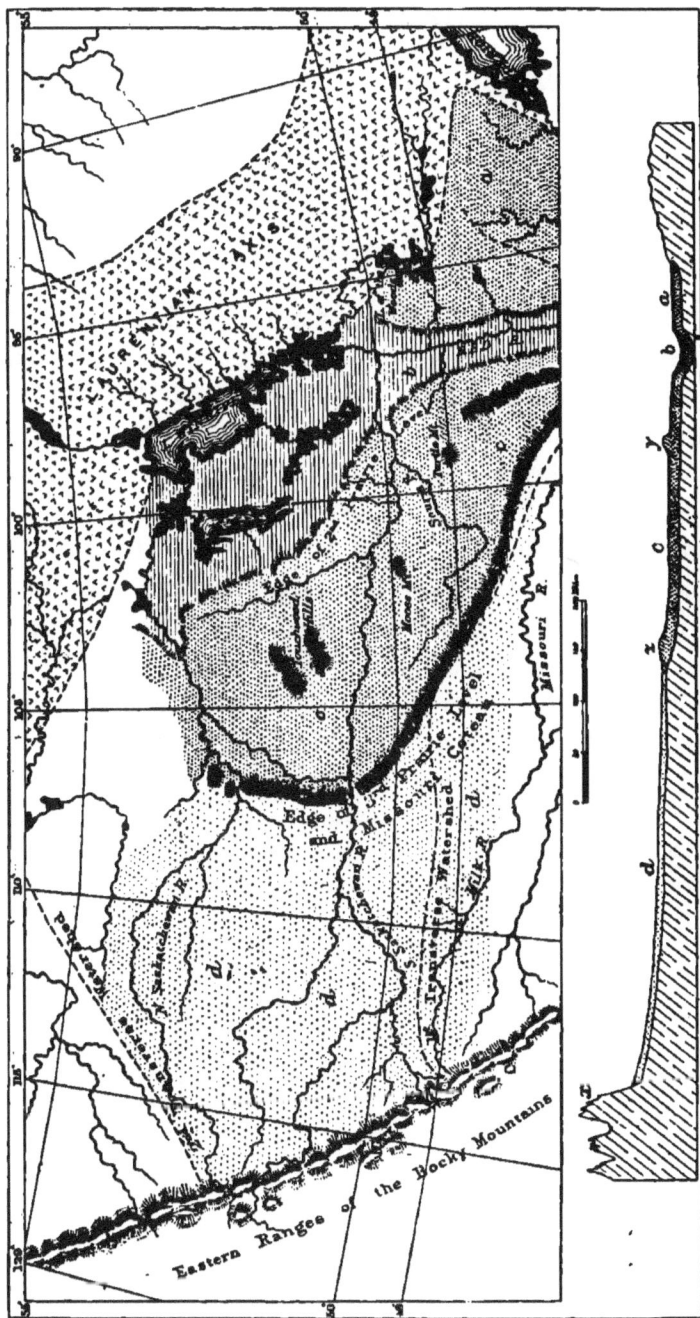

Fig. 7.—*Map and section showing the distribution of Drift on the western plains.* (a) Boulder-drift east of Red river; (b) Lacustrine deposit of Red River valley; (c) Boulder-drift of second Prairie level; (d) Boulder-drift of higher Prairie; (x, y) Missouri coteau and outliers; (z) Rocky mountains. (After G. M. Dawson.)

CHAPTER IV.

PHYSICAL AND CLIMATAL CONDITIONS
(CONTINUED).

II.—Causes of Glaciation and Distribution of Erratics—Continued.

2.—SEA-BORNE ICE.

2. We now come to the question of dispersion of boulders and formation of boulder-clay and striated surfaces by floating ice, whether formed on the sea or derived from the ends of glaciers discharging at the level of the sea. Here we may take for granted the great submergence of the North American continent, first in the early glacial period, and subsequently to a still greater extent in the later glacial period, and that this submergence was in the earlier period differential, affecting the plains and not the mountains.

I shall first quote here a letter received several years ago from my friend, Dr. John Rae, F.R.S., a traveller of almost unrivalled arctic experience,* and which shows

---

* September, 1882, The same facts are referred to in a paper by Dr. Rae in the Journal of the Physical Society, 1881.

very clearly what may be effected by floating ice impinging on shores :

"Having learnt that you took much interest in the transportation of boulders by ice, I venture to mention a peculiar mode in which this is done, which I have not seen noticed elsewhere.

"When at Repulse bay in 1846–7, I noticed in the spring that large boulders, some of them more than three feet in diameter, gradually appeared on the surface of the sea-ice near shore, as the ice wasted away by thaw and evaporation. Although I wondered how this was done, I had no opportunity of proving it until wintering at the same place in 1853–4.

"First let me say that the rise and fall of the tide is from eight to ten feet, sometimes more, and that the ice on Repulse bay attains a thickness of about eight feet.

"Suppose, then, that a boulder of a good size, say $3\frac{1}{2}$ feet diameter, is lying on the shore, at, or farther out than low-water mark, in the beginning of winter, when the ice is forming, it comes in contact with the boulder at low water, and the boulder breaks through it; but when the ice becomes two feet or so thick, the boulder freezes to it, and is lifted with the ice when the tide rises. This continues all winter, the ice increasing in thickness to eight feet, soon encloses the boulder *inside* of it, and having about four feet of solid ice below. Boulders may thus be carried hundreds of miles.

"There is another phase in connection with this matter. Supposing boulders are lying in five or six feet of water or less when it is low water, these boulders would get frozen to the lower surface of the ice, and get set into it as a diamond is set for cutting glass, and would thus be a good graver of any rocks it might pass over."

To the same effect is the following by Capt. Fielden: *

"*Sea-ice*, moved up and down by tidal action, or driven on shore by gales, was found to be a very *potent agent* in the *glaciation of rocks and pebbles.* The work was seen in progress along the shores of the Polar Basin. At the south end of a small island in Blackcliff bay, lat. 82° 30′ N., the bottoms of the hummocks, some eight to fifteen feet thick, were studded with hard limestone pebbles, which, when extracted from the ice, were found to be *rounded and scratched* on the exposed surface only.

"On shelving shores, as the tide recedes, the hummocks, sliding over the subjacent material down to a position of rest, make a well-marked and peculiar sound, resulting from the *grating of included pebbles, with the rocky floor beneath,* or in some cases on other pebbles included in drift overlying the rock."

Action of this kind now taking place along northern shores must have been carried over all the submerged portions of the continents in the Pleistocene, and affords the only rational mode of accounting for the general striation, not dependant on local glaciers and related to the lines of valleys, but occurring on the surfaces of the plains, and on the summits exposed at various times to the action of ice carried by the northern currents.

Large boulder of sandstone deposited by modern ice on a sand-bank— Petitcodiac river, New Brunswick.

---

* Nares' Arctic Voyage, Vol, II., p. 343, quoted, along with other examples, by Mr. Milne Home.

I have already, in Chapter II., referred to the modern boulder-belt of the shores of the estuary of the St. Lawrence as seen at Little Metis, and may now adduce it as an example of a pseudo-moraine as well as of detached groups of boulders, produced by the present field-ice in winter, and giving proof of the movement of great boulders by this agency and the piling of them up, at and near the line of low water. The frontispiece shows the appearance of the boulder-belt at Little Metis, and it is to be borne in mind that the greater part of the boulders were originally derived from the Laurentian hills thirty-five miles or more distant to the northward, and that these boulders are moved about from year to year by the ice. Similar facts collected by Capt. Bayfield will be found in Lyell's Principles of Geology.

I should add here that while the old glacial striae on the rocks at Little Metis run N.N.E., there are on the modern pavements of boulders less pronounced striations running in a similar direction, or about N.E., and which must be produced by the modern field-ice drifting up and down with the tidal currents.

We are indebted to Prof. H. Y. Hind for a graphic description of the action of sea-ice on the coast of Labrador, in the form known as " Pan Ice." *

" ' Pan ice ' is derived from bay ice, floes, and coast ice, varying from five to ten or twelve feet in thickness, all of which are broken up during spring storms. When the disruption of the ice-sheet which seals the fiords, the island zone, and the sea itself for many miles outside, continuously, is effected in June, the resulting ' pans,' as the fishermen term them, vary in size from a few square

---

* Canadian Naturalist, Vol. VIII., 1877.

yards to many acres in extent. The uniform and unbroken mass of ice in the winter months has no lateral motion; it rises and falls with the tide, but is unaffected by winds until the warmth of spring softens its hold on the islands to which it is keyed. When the pans are pressed on the coast by winds, they accommodate themselves to all the sinuosities of the shore-line, and being pushed by the unfailing arctic current, which brings down a constant supply of floe ice, the pans rise over all the low-lying parts of the islands, grinding and polishing exposed shores, and rasping those that are steep-to. The pans are shoved over the flat surfaces of the islands, and remove with irresistible force every obstacle which opposes their thrust, for the attacks are constantly renewed by the ceaseless ice-stream from the north-east, and this goes on uninterruptedly for a month or more. Sometimes a change in the wind brings the endless sheet back again, and it is the middle of July before some of the fiords are clear of ice. Hence boulders, shingle, and beaches are rarely seen except in sheltered nooks and coves, and the masses, *pushed* or torn from those surfaces where cleavage offers a chance of disruption, are urged into the sea and rounded into boulder form by the rasping and polishing pans.

"But this is not all of the work of pan ice. The bottom of the sea, to the depth of twelve or fifteen feet, and at all less depths, is smoothed and planed by the drifting masses when they pile one on the other, and at depths less than eight feet, when the pans are driven before the wind or carried by the currents. In sailing from Aillik to Nain or to Cape Mugford, the fishermen send a man aloft to look out for 'White Rocks.' These are prominences or swells in the general level of the sea-bottom among the islands, from which every particle of sea-weed has been removed by pan ice.

"During a period of subsidence, the blocks of stone, boulders, mud, and sand, pushed to and fro on the shallow sea-bottom by pan ice, ultimately accumulate in hollows and ravines below its action; and when the debris is pushed into profound submarine valleys, such as exist on the Labrador coast, the mass will resemble boulder-clays, and in a sinking marine area it will accumulate to a great thickness; in a rising area it would be liable to be remodelled by the action of the waves, except in the case of very deep valleys. There are not many known narrow and profound submarine valleys on the north-eastern coast of Labrador, but those which are known offer precisely the conditions required for the accumulation of boulder-clays or drift by the action of pan ice.

" The seaward extension of Uksuktak fiord, which lies a little to the south of Hopedale, affords an apt illustration. Commander Maxwell's soundings show a profound sub-marine ravine between clusters of islands for upwards of eight miles, in which the depth reaches 124, 126, 123, 106, and 130 fathoms. Between the islands of Niatak and Paul, near Nain, the lead shows 71 fathoms. It is evident that the material torn from the surrounding islands by pan ice, and pushed along the bottom of the sea into these profound submarine valleys during a period of general submergence, will be protected from the action of the waves, and the loose blocks and boulders will have a forced arrangement in the mud, as if they had been pushed over a bank, and thus produce the irregular dis-position so frequently seen in boulder-clay deposits. In such narrow and profound valleys as those instanced, the accumulation of boulder-drift probably goes on at the present time, and may continue during a period of elevation, until large portions of the drift are raised above the sea-level and beyond the influence of the waves, which will attack only its sea front. But the agent which gives rise to this heterogeneous mass is pan ice, and the formation of boulder clay is very probably a part of its work over a vast area on the Labrador coast at the present day, throughout the labyrinth of islands which fringe that coast to a depth of 20 miles seawards."

There can be no doubt, as Dr. G. M. Dawson has pointed out in the case of the western plains and of British Columbia, that the typical boulder-clay spread over the country and filling pre-existing hollows is much more of the nature of the deposits now forming by field, floe, and pan ice under water, than of anything of the nature of the bottom moraine of a glacier. Its material may, however, have been added to and its arrangement affected by the bergs thrown off from the foot of glaciers terminating in the sea.

An important question arises here as to the means of distinguishing sea from land glaciation. This I believe to be quite possible if careful observations are made. Sea glaciation is always accompanied with much smoothing and polishing, and on very hard rocks the striation is comparatively imperfect, while it is usually not quite uniform in direction and often presents two sets of striae. The action of true land glaciers, especially when thick and moving down considerable slopes, produces deep grooves, as well as striae, on vertical as well as horizontal surfaces, and is more fixed and uniform. The more intense forms of sea glaciation, especially if long continued, may, however, approach very closely to the effects of land ice.

The action of icebergs is undoubtedly, though not the chief, one of the most important manifestations of ice-power; and while it is quite wrong to designate any theory of glaciation by floating ice as an "Iceberg theory," these huge ice-islands are not to be neglected in our estimate of factors in the work of transport and planing of surfaces. Their main agency is, of course, in the arctic seas, but their effects are felt as far south as the coast of Newfoundland and the entrance to the gulf of St. Lawrence, where alone I have had any opportuity of observing them.

The snow-clad hills of Greenland send down to the sea great glaciers, which, in the bays and fiords of that inhospitable region, form at their extremities huge cliffs of solid ice, and annually "calve," as the seamen say, or give off a great progeny of ice islands which, slowly drifted to the southward by the arctic current, pass along the American coast, diffusing a cold and bleak atmosphere, until they melt in the warm waters of the Gulf stream. Many of these bergs enter the straits of Belle-Isle, for the arctic current clings closely to the coast, and a part of it seems to be deflected into the gulf of St. Lawrence through this passage, carrying with it many large bergs.

Mr. Vaughan, late superintendent of the lighthouse at Belle-Isle, has kept a register of icebergs for several years. He states that for ten which enter the straits, fifty drift to the southward, and that most of those which enter pass inward on the north side of the island, drift toward the western end of the straits, and then pass out on the south of the island, so that the straits seem to be merely a sort of eddy in the course of the bergs. The number in the straits varies much in different seasons of the year. The greatest number are seen in spring, especially in May and June; and toward autumn and in the winter very few

remain. Those which remain until autumn are reduced
to mere skeletons; but if they survive until winter, they
again grow in dimensions, owing to the accumulation
upon them of snow and new ice. Those that I have seen
early in July were large and massive in their proportions.
The few that remained in September were smaller in size
and cut into fantastic and toppling pinnacles. Vaughan
records that on the 30th of May, 1858, he counted in the
straits of Belle-Isle 496 bergs, the least of them sixty feet
in height, some of them half a mile long and two hundred
feet high. Only one-eighth of the volume of floating ice
appears above water, and many of these great bergs may
thus touch the ground in a depth of thirty fathoms or
more, so that if we imagine four hundred of them moving
up and down under the influence of the current, oscillating
slowly with the motion of the sea, and grinding on the
rocks and stone-covered bottom at all depths from the
centre of the channel, we may form some conception of
the effects of these huge polishers of the sea-floor.

Of the bergs which pass outside of the straits, many
ground on the banks off Belle-Isle. Vaughan has seen a
hundred large bergs aground at one time on the banks,
and they ground on various parts of the banks of Newfoundland, and all along the coast of that island. As
they are borne by the deep-seated cold current, and are
scarcely at all affected by the wind, they move somewhat
uniformly in a direction from N.E. to S.W., and when
they touch the bottom the striation or grooving which
they produce must be in that direction.

In passing through the straits in July, one sees a great
number of bergs. Some are low and flat-topped with
perpendicular sides, others concave or roof-shaped like
great tents pitched on the sea; others are rounded

9

in outline or rise into towers and pinnacles. Most of
them are of a pure dead white, like loaf sugar, shaded
with pale bluish green in the great rents and recent
fractures. A few of them seem as if they had grounded
and then overturned, presenting a flat and scored surface
covered with sand and earthy matter.

Viewed as geological agents, the icebergs are, in the first
place, parts of the cosmical arrangements for equalizing
temperature, and for dispersing the great accumulations
of ice in the arctic regions, which might otherwise
unsettle the climatic and even the static equilibrium of
our globe, as they are believed by some imaginative
physicists and geologists to have done in the so-called
glacial period. If the ice-islands in the Atlantic, like
lumps of ice in a pitcher of water, chill our climate in
spring, they are at the same time agents in preventing a
still more serious secular chilling which might result
from the growth without limit of the arctic snow and ice.
They are also constantly employed in wearing down the
arctic land, and aided by the great northern current from
Davis's straits, in scattering its debris of stones, boulders
and sand over the banks along the American coast.
Incidentally to this work, they smooth and level the
higher parts of the sea bottom, and mark it with furrows
and striae indicative of the direction of their own motion.
In this manner multitudes of boulders from Baffin's bay
are annually distributed along the bed of the arctic
current off the American coast, and are buried in the
accumulations of mud which are being laid down on the
banks by this current; while in the strait of Belle-Isle
the same effects are being produced, on a small scale,
which, in the Pleistocene period, were produced in the
greater and wider strait then formed by the St. Lawrence

valley, and in which the icebergs from the far north were probably reinforced by great numbers of similar masses descending from the Laurentian hills on the north side of that valley, as well as by the field-ice formed along its shores.

I have referred in Acadian Geology * to the ingenious theory of Darwin as to the transport of boulders from lower to higher levels by floating ice in a subsiding condition of the land.† This theory, in my judgment, still affords the only satisfactory explanation of such facts as the transference of slabs of sandstone from the plains of Cumberland and the St. Mary's river in Nova Scotia to the summits of hills several hundred feet higher than the original seats of the erratics. Facts of this kind are not infrequent throughout Eastern Canada, and are quite inexplicable on any theory of land glaciation.

As to transport of materials by floating ice, it is almost superfluous to give farther details. A few examples and a few applications to the Pleistocene may be mentioned. We have already seen that extensive boulder-drift is now taking place in the lower St. Lawrence, and that our boulder beaches and pavements almost rival the so-called moraines of the Pleistocene. Even on lake margins the ice produces appearances of the same kind on a small scale. The writer long ago described these in Nova Scotia,‡ and Spencer has correlated the ancient and modern margins on the larger Canadian lakes.§ The removal of large boulders by the ice is a matter of constant occurrence on our shores, and the dredges of the "Challenger" took up

---

* Fourth Edition, p. 65.
† Journal of London Geol. Society, Vol. IV., p. 315.
‡ Acadian Geology. Report on Prince Edward Island.
§ Bull. Geol. Socy. America, Vol. I.

boulders from the banks off the American coast, from which I had previously recorded travelled stones taken up by the hooks of fishermen which became fixed on organisms growing on them. Off the ends of the Greenland glaciers in Baffin's bay and elsewhere, such deposits must be proceeding on a gigantic scale. The Reports of the "Challenger" show, as already stated, that over vast oceanic areas lying to the north of the antarctic continent, deposits of stones and other debris falling from ice are so abundant as to mask the organic accumulations. In like manner immense deposits of submarine inorganic matter are being deposited in the arctic seas in the track of the icebergs and the drift floe-ice.

If now we turn to the Pleistocene accumulations on the land, I have shown that throughout the valley of the lower St. Lawrence the old till or boulder-clay contains marine shells, and in the overlying deposits, the upper Leda clay and Saxicava sand, these are extremely abundant. Both of these deposits contain far-travelled boulders often of great size, and these have been carried to great heights. On Montreal mountain marine shells occur at an elevation of nearly 600 feet, and at a still greater height boulders which have been derived from the Laurentian highlands to the north. On still higher terraces, up to 1,200 feet, from Labrador * to the foot of lake Ontario, there are shore beaches and boulders, though in the west they have not afforded marine shells.

To the southward, Upham has found marine shells in the boulder-clay near Boston up to an elevation of 200 feet.† It is true this is in Drumlins or detached hills,

* Richardson.
† Am. Journal of Science, May, 1889.

but these are not unlikely undenuded portions of former beds. He also supposes them to have been pushed up from the sea by an ice-sheet, which, however, I am sure, if consulted, would refuse to do him any such service.

Dr. Bell informs me that drift deposits containing shells occur on the north of the Laurentian axis, facing Hudson's bay, in many localities, and that in one of these they reach to within 133 miles of lake Superior, and are at an elevation of 625 feet, or very nearly that of the lake itself.

I have already referred to the observations of Dr. G. M. Dawson with reference to the Missouri coteau, one of the greatest ridges of drift in the world. His description of it merits quotation here, as a remarkable example of an old sea margin.*

"The great drift-ridge of the Missouri coteau at first sight resembles a gigantic glacier-moraine; and, marking its course on the map, it might be argued that the nearly parallel line of elevations, of which Turtle mountain forms one, are remnants of a second line of moraine produced as a feebler effort by the retiring ice-sheet.

"Such a glacier must either have been the southern extension of a polar ice-cap, or derived from the elevated Laurentian region to the east and north; but I think, in view of the physical features of the country, neither of these theories can be sustained.

"To reach the country in the vicinity of the forty-ninth parallel a northern ice-sheet would have to move up the long slope from the Arctic ocean and cross the second transverse watershed; then, after descending to the level of the Saskatchewan valley, again to ascend the

* Quarterly Journal London Geol. Society, 1875.

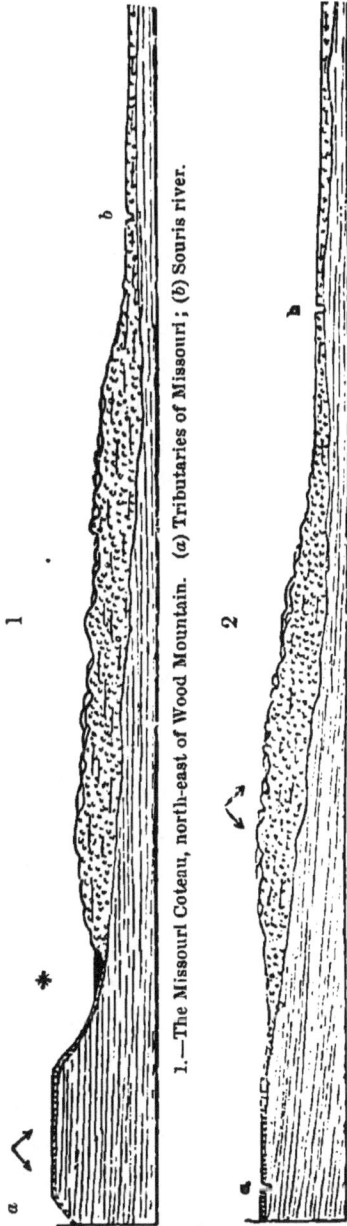

1.—The Missouri Coteau, north-east of Wood Mountain. (a) Tributaries of Missouri; (b) Souris river.

2.—The Missouri Coteau, near the 49th parallel. Letters as before. East of the Coteau the prevalent drift is from the north-east. West of the Coteau it is from the westward. (After Dr. G. M. Dawson.)

slope (amounting, as has been shown, to over four feet per mile) to the first transverse watershed and plateau of the Lignite Tertiary. Such an ice-sheet, moving throughout on broad plains of soft, unconsolidated Cretaceous and Tertiary rocks, would be expected to mark the surface with broad flutings parallel to its direction, and to obliterate the transverse watersheds and valleys.

" If it be supposed that a huge glacier, resting on the Laurentian axis, spread westward across the plains, the physical difficulties are even more serious. The ice moving southward, after having descended into the Red-River trough, would have had to ascend the eastern escarpment of soft Cretaceous rocks forming its western side, which in one place rises over 900 feet above it. Having gained the second prairie-steppe, it would have had to pass westward up its sloping surface, surmount the soft edge of the third steppe without much altering its form, and finally terminate over 700 miles from its source, and at a height exceeding the present elevation of the Laurentian axis by over 2,000 feet. The distribution of the drift equally negatives either of these theories, which would suppose the passage of an immense glacier across the plains.

"In attributing the glacial phenomena of the great plain to the action of floating ice, I find myself in accord with Dr. Hector, who has studied a great part of the basin of the Saskatchewan—and also, as far as I can judge from his reports, with Dr. Hayden, who, more than any other geologist, has had the opportunity of becoming familiar with all parts of the Western States.

"The glaciating agent of the Laurentian plateau in the Lake of the Woods region, however, cannot have been other than glacier ice. The rounding, striation and polishing of the rocks there are glacier work; and icebergs floating, with however steady a current, cannot be supposed to have passed over the higher region of the watershed to the north, and then, following the direction of the striæ and gaining ever deeper water, to have borne down on the subjacent rocks. The slope of the axis, however, is too small to account for the spontaneous descent of ordinary glaciers. In a distance of about 30 miles, in the vicinity of the Lake of the Woods, the fall of the general surface of the country is only about 3½ feet to the mile. The height of the watershed region northeast of the lake has not been actually measured; but near Lac Seul, which closely corresponds with the direction required by glaciation, according to Mr. Selwyn's measurements, it cannot be over 1,400 feet. The height of land

in other parts of the Laurentian region is very uniformly
between about 1,600 and 1,200 feet. Allowing, then,
1,600 feet as a maximum for the region north-east of the
Lake of the Woods, and taking into account the height of
that lake and the distance, the general slope is not greater
than about three feet per mile—an estimate agreeing
closely with the last, which is for a smaller area and
obtained in a different way. This slope cannot be considered sufficient to impel a glacier over a rocky surface,
which Sir William Logan has well characterized as
'mamillated,' unless the glacier be a confluent one pressed
outwards mainly by its own weight and mass.

"Such a glacier, I conceive, must have occupied the
Laurentian highlands ; and from its wall-like front were
detached the icebergs which strewed the *débris* over the
then submerged plains, and gave rise to the various
monuments of its action now found there.

"The sea, or a body of water in communication with it,
which may have been during the first stages of the
depression partly or almost entirely fresh, crept slowly
upward and spread westward across the plains, carrying
with it icebergs from the east and north. During its
progress most of the features of the glacial deposits were
impressed. In the section described at Long river, we
find evidence of shallow current-deposited banks of local
material, afterwards, with deepening water, planed off by
heavy ice depositing travelled boulders.

"The sea reaching the edge of the slope constituting
the front of the highest prairie-level, the deposition of the
Coteau began, and must have kept pace with the increasing depth of the water, and prevented the action of heavy
ice on the front of the Tertiary plateau. The water may

also have been too much encumbered with ice to allow the formation of heavy waves.

"The isolated drift highlands of the second plateau, including the Touchwood hills, Moose mountain and Turtle mountain, must also at this time have been formed. With regard to the two former, I do not know whether there is any preglacial nucleus round which drift-bearing icebergs may have gathered. There is no reason to suppose that Turtle mountain had any such predisposing cause ; but it would appear that a shoal once formed, by currents or otherwise, must have been perpetuated and built up in an increasing ratio by the grounding of the floating ice.

"The Rocky mountains were probably also at this time covered with descending glaciers ; but these would appear to have been smaller than those of the Laurentian axis, as might, indeed, be pre-supposed from their position and comparatively small gathering-surface. The sea, when it reached their base, received from them smaller icebergs ; and by these and the shore-ice the *quartzite-drift* deposits appear to have been spread. That this material should have travelled in an opposite direction to the greater mass of the drift is not strange; for while the larger eastern and northern icebergs may have moved with the deeper currents, the smaller western ice may have taken directions caused by surface-currents from the south and west, or even been impelled by the prevailing winds. Some of the Laurentian *débris*, as we have seen, reached almost to the mountains, while some of the *quartzite-drift* can be distinguished far out towards the Laurentian axis.

"The occurrence of Laurentian fragments at a stage in the subsidence when, making every allowance for subsequent degradation, the Laurentian axis must have been far below water, would tend to show that the weight and mass of the ice-cap was such as to enable it to remain as a glacier till submergence was very deep.

"The emergence of the land would seem to have been more rapid; or, at least, I do not find any phenomena requiring long action at this period. The water in retreat must have rearranged to some extent a part of the surface-materials. The quartzite-drift of the third steppe was probably more uniformly spread at this time, and a part of the surface sculpture of the drift-deposits of the second plateau may have been produced. It seems certain, however, that the Rocky mountains still held comparatively small glaciers, and that the Laurentian region on its emergence was again clad to some extent with ice, for at least a short time. The closing episode of the glacial period in this region was the formation of the great fresh-water lake of the Red River valley, or first prairie-level (which was only gradually drained), and the re-excavation of the river-courses.

"It must not be concealed that there are difficulties yet unaccounted for by the theory of the glaciation and deposit of drift on the plains by icebergs; and chief among these is the absence, wherever I have examined the deposits and elsewhere over the West, of the remains of marine mollusca or other forms of marine life. With a submergence as great as that necessitated by the facts, it is impossible to explain the exclusion of the sea; for, besides the evidence of the higher western plains and Rocky mountains, there are terraces between the Lake of the Woods and lake Superior nearly to the summit of the Laurentian axis, and corresponding beach-marks on the face of the northern part of the second prairie escarpment.

"Mr. Belt, in an interesting paper (Quart. Journ. Geol. Soc., Nov., 1874), deals with similar difficulties in explaining the glaciation of Siberia. The northern part of Asia appears in many ways to resemble that of America; surrounded by mountain-chains on all sides save the north, it is a sort of interior continental basin covered with 'vast level sheets of sand and loam.' As in the interior regions of America, marine shells are absent, or are only found along the low ground of the northern coast. To account for these facts, Mr. Belt resorts to a theory first suggested by him eight years ago, by which he supposes the existence of a polar ice-sheet capable of blocking up the entire northern front of the country, and damming back its waters to form an immense fresh-water lake. The outfall of this lake, during its highest stage, he supposes to have been through the depression between the southern termination of the Ourals and the western end of the Altai to the Aral and Caspian seas."

The main difficulty in the way of this masterly explanation is the great height above the sea of the western part of the plains; but this is now met by the probability of the depression of the plains contemporaneously with the elevation of the Cordillera, since suggested by the author of the extract. To the absence of marine shells from the deposits of the plains no importance need be attached. The water may have been cold and brackish, and in all geological periods gravels, sands and conglomerates usually have few marine fossils.

In 1883 I had an opportunity of going over the same ground, and my notes respecting it are as follows: *

The Great Missouri coteau to which Dr. G. M. Dawson first directed prominent attention as a glacial

* See Journal Geol. Society of London, 1883.

feature, and which fringes the margin of the third plateau,
about 400 miles west of Winnipeg, is now known to be
continuous with similar ridges extending southward into
the United States and eastward towards the Atlantic,
and which have been described as the terminal moraine
of a great continental glacier. In the western plains,
however, where it has its greatest development, it cannot
be explained in this way, but must mark the margin of
an ancient glacial sea, or at least of that deeper portion of
such sea in which heavy ice could float, while in its upper
portion it shows evidence of having been, in the later
periods of its formation, an actual water-margin.

The railway, taking advantage of the oblique valley of
Thunder creek, crosses the coteau at one of its least-
marked portions, but where it still presents very definite
and striking characters. On entering it, the railway
passes for nearly thirty miles through a rolling or broken
country, consisting of successive ridges and mounds inter-
spersed with swales and alkaline ponds without outlet.
To this class belongs a somewhat extensive series of lakes
known as the "Old Wives' Lakes." The highest point of
the coteau on this section is near Secretan Station.

As seen in the road-cutting, the basis of the ridges
appears to consist of thick beds of imperfectly stratified
clay, derived from the disintegration of the local Cretaceous beds, but with many Laurentian boulders. In one
place the clay was observed to be crumpled as if by
lateral pressure. Above the clay are stratified gravels,
also with large boulders, most abundant at top. The
ridges are highest and most distinct at the eastern or
lower side, and gradually diminish towards the upper or
western margin, where they terminate on the broadly
rolling surface of the upper prairie.

The history of the coteau would seem to have been as follows :

1. The excavation in pre-glacial times of an edge or escarpment in the gently sloping surface of the Cretaceous and Laramie beds, and the cutting by subaerial causes of coulées and valleys of streams in this escarpment.

2. Submergence in the glacial period, in such a manner as to permit heavy ice loaded with Laurentian *débris* to ground on the edge of the escarpment and deposit its burden there, while at the period of greatest submergence deep water must have extended much further westward. These conditions must have continued for a long time and with somewhat variable depth of water.

3. Re-elevation, during which gravel ridges were formed, until at length the coteau became the coast-line of a shallow sea, which lingered at a later date along the line already referred to in advance of the coteau.

4. On the re-elevation of the country, the transverse ravines and valleys were so effectually dammed up by the glacial ridge, that the surface waters of the region, now comparatively arid, have to remain as alkaline lakes and ponds behind the coteau.

The upper prairie plateau, extending from the coteau to the Rocky mountains, has, on its general surface, comparatively few boulders ; yet these are locally numerous, especially on the eastern and northern sides of some gentle elevations of the prairie. They consist, as before, of Laurentian gneiss, Huronian schists, and yellow Silurian limestone, all derived from the eastern side of the plains, some of the boulders of Laurentian gneiss being of great dimensions. Some of these have been used in modern times by the buffalo as rubbing-stones, and are surrounded by basin-shaped depressions formed by the feet of these animals.

That strong currents of water have traversed this upper plain, is shown not only by the occasional ridges of gravel, but by the depressions known as "slues," which must have been excavated subaqueous currents.

Near Medicine Hat a terrace of boulders was seen at an elevation of about 200 feet above the river; and in sections of the drift observed in coulées, the boulders were seen to be arranged in layers; but whether these appearances had relation to fluviatile action, before the excavation of the deep valley of the Saskatchewan, or belonged to the orignal distribution of the drift, was not apparent.

Laurentian boulders were seen all the way to Calgary, but with an increasing proportion of quartzite boulders from the Rocky mountains; and on the banks of the Bow river were large beds of rounded pebbles which must have been swept by water out of the valleys of the mountains, and are quite similar to those now observed in the bed of the Bow itself.

Beyond this, Dr. G. M. Dawson has recorded Laurentian boulders and fragments of limestone from the eastern Palæozoic beds, at elevations of from 4,200 to above 5,000 feet,* at the foot of the Rocky mountains, evidencing a driftage of at least 800 miles, and an elevation considerably above that of the sources from which they came. He well observes that anything which would explain the

---

* "Many of these (Laurentian erratics along or near the base of the mountains between the 49th and 50th parallels) lie at heights exceeding 4,000 feet, while the highest observed instances of their occurrence are at an elevation of 5,289 feet above the present sea-level, the erratics being here stranded upon moraine ridges due to local glaciers which have flowed out from the valleys of the Rocky mountains, probably during the first maximum of glaciation. These erratics are known to have come a distance of at least 500 miles from the eastward."—G. M. Dawson.

origin of the coteau must also explain the transport of these boulders so far above it and beyond its limits, as well as the contemporaneous distribution of boulders from the Rocky mountains to the eastward. These phenomena are explicable on the hypothesis of a glacial sea of varying depth, but not on that of land glaciation, which would also be inapplicable in a region necessarily of so small precipitation of moisture and occupied by soft deposits so little suited to the movement of glaciers. *A fortiori* the same explanation applies to that great tail of *débris* extending from the southern end of the Missouri coteau across the continent, and which forms the great "terminal moraine" of the continental glacialists. The fact that this so-called moraine sometimes occurs where there is no elevated shore immediately outside of it constitutes no objection to this, since there may have been unequal elevation. There is, nevertheless, good evidence of the action of glaciers on a large scale in certain portions of the glacial periods, both on the Rocky mountains and on the Laurentian hills and table-lands to the east.

3.—ICE-FRESHETS IN RIVERS.

3. A cause of boulder-drift to which too little importance has been attached, is what may be termed "ice freshets" in the rivers of northern latitudes. Lyell has summed up some facts of this kind in relation to the rivers of Siberia, and Belgrand has referred to the evidence in the valley of the Somme. On a small scale, I have noted the effects of these ice-floods in Nova Scotia and New Brunswick. They occur in early spring, when sudden thaws and violent rains sometimes occur before the ice in the rivers has broken up. In these circumstances, the rivers rising break up the ice on their

surfaces, and sweep it downward, laden with uprooted trees, timber, stones and gravel. The destruction of roads, bridges, and other property, and the tearing up and burying under rubbish of meadows, are sometimes terrific. Fortunately, such freshets occur only at long intervals, but the loss and injury which they cause are long remembered, and the ridges and mounds of *débris* which they deposit remain as mementoes of their destructive power. Logan has well described* the annual breaking up, or "shove," of the ice on the St. Lawrence, which, though a comparatively quiet phenomenon, piles up ridges of stone where the floes of ice ground. In the Pleistocene period, such ice-freshets and shoves must have been frequent, and it is not unlikely that some of the gravel deposits which are credited to the melting of the continental glacier are due to their agency.

4.—BORDAGE ICE.

4. A special ice agency of some importance is that to which Mr. Chalmers has directed attention on the coast of the bay des Chaleurs.†

Mr. Chalmers describes the rocks of various paleozoic periods, along the south side of the bay des Chaleurs, as presenting a somewhat flat and even surface to a height of 50 to 75 feet above the sea level. A similar appearance is presented by the beds below the sea level along the coasts. He connects this with the action of floating ice, now very evident in the bay. In winter a fixed border of ice is formed along the coast, from two to six feet thick, and extending from the shore for a distance of from half a mile to several miles. The open portion of the bay is generally full of loose floes.

* Journal Geol. Society.
† Canadian Record of Science.

In March and April the marginal sheets break up into floes, and drift up and down the bay, and the ice in the bay is often reinforced by large fields from the gulf without. These sheets of ice grind over the reefs and impinge on the shores with great force, and, evidently, at present, exert a great erosive and transporting power. In the latter part of the Pleistocene period, when the land stood at a lower level and the climate was, possibly, colder, their action may have been still more powerful. This action of floating ice is similar to that which has been pointed out by Admiral Bayfield in the river St. Lawrence, and by the writer on the coast of Nova Scotia; but Mr. Chalmers believes that it has had a somewhat exceptional power on the south side of the bay des Chaleurs, which renders its influence there unusually conspicuous and instructive.

5.—ICE IN TIDAL ESTUARIES.

5. Still another form of ice-drift is that of ice-floes in tidal estuaries, which is seen in, perhaps, its extreme development in those of the bay of Fundy. In Acadian geology I have noticed the removal of large boulders in this way, and the Lower St. Lawrence may be regarded as a tidal estuary; but I have seen merely the effects, not the actual operation, of the ice in winter and early spring, and Hind has given so graphic and complete a picture of the phenomena,* that I cannot do better than reproduce it in his own words. The agency which he describes has, not improbably, been concerned in the production of those curious patches of sand and clay frequently seen in boulder-clay and gravel beds, and whose origin is often difficult to comprehend.

---

* *Canadian Monthly*, Sept. 1875.

"The appearance of an estuary in the bay of Fundy at any time in mid-winter, presents some singular and striking phenomena, which may contribute to our knowledge of the manner in which different agents have assisted in excavating this extraordinary bay, and are now engaged in extending its domain in some directions and reducing it in others.

"Within an hour or so of flood tide, the estuary is seen to be full of masses of floating ice, mud-stained, and, sometimes, but not often, loaded with earth, stones, or pieces of marsh. The tide, flowing at a rate of four or five miles an hour, rushes past with its broad, ice-laden current until the flood. A rest, or 'stand,' then occurs, of variable duration. During this brief period all is repose and quiet, but as soon as the ebb begins, the innumerable blocks of ice commence to move, and in half an hour they are as swiftly gliding noiselessly towards the sea, as an hour before they swiftly and noiselessly glided from it. It produces in the mind of one who sees these ice-streams for the first time, moving up the wide river faster than he can conveniently walk, a feeling of astonishment, akin to awe, which is heightened rather than diminished if he should return to the same point of view half an hour later, and find the ice-stream rushing as impetuously as before in exactly the opposite direction.

"During the ebb tide many of the larger blocks ground on the sand-bars, so that when the tide is out the extensive flats are covered with ice-blocks innumerable. If the period between the ebb and the return of the flood is very cold, the stranded ice-blocks freeze to the sand-bars or mud-flats and are covered by the returning tide, but only until the warm tidal water succeeds in thawing the frozen sand or mud around the base of the ice-block, and

it is enabled, by means of its less specific gravity, to
break away with a frozen layer of mud or sand attached
to it. It reaches the surface of the water with a bound,
and is instantly swept away by the incoming tide. The
spectacle thus presented by an extensive sand-bar after a
few hours of freezing weather, is most extraordinary;
the whole surface of the flood or ebb becomes suddenly
alive with blocks of ice, springing up from below, each
carrying away its burden of sand or mud frozen to its
base. Later in the season, towards the middle of March,
this singular phenomenon can be seen to the best advantage,
and it is curious to watch a block of, say, ten feet square
by five or six in thickness, being gradually covered by the
tide until it becomes lost to view for an hour or more,
during which time the water may have risen three or four
feet above it. 'When least expected' up the submerged
mass springs; it has broken loose from the frozen bottom,
it seems to stagger and pause for a few moments at the
surface, and then joins the rest of the icy stream on their
monotonous journey, until it is again stranded on some
other flat or bar during the ebbing tide. But this is only
a small part of the history of these ice-blocks, for, during
neap tides, it often happens that a block is stranded in
such shallow water that the flood has not power to raise
it from the substratum to which it is frozen. The block
grows there with every tide; fresh films of ice and tidal
mud form all round it four times during every twenty-
four hours. It receives accessions from falling snows, and,
by the time the spring tides begin, it has greatly increased
in size and is more firmly frozen or weighted to the sand-
bar. Even the spring tides may not have the power to
free it from its icy bonds if the weather has been
extremely cold; the consequence is that it goes on

increasing in size, and actually becomes a miniature berg, containing some thousands of cubic feet of ice and mud, and still retaining a buoyancy which will enable it, after a thaw, during high spring tides, to break away with a load of *débris,* and carry it either out to sea or up the estuary, and, if it should chance to be stranded again, it will probably leave a portion of its burden, provided it has not melted off during its voyage with the tide. But there can be no doubt that some of the attached sand, mud, or shingle is melted off during the journey of the block or miniature berg, and drops into the bed of the river or estuary. In reality, these ice-cakes, when in motion, are perpetually strewing the bottom with transported material and bringing a portion from one place to another, during about five hours of the flood, and carrying part of it back again, during five hours of ebb, to the limits of the backward and forward tidal range of each particular ice-cake. But when they accumulate in an eddy, they become powerful carriers and depositors of detritus, and if artificial obstructions be introduced so as to form an eddy in the usual course of the ice-stream, the accumulation must necessarily be very rapid."

6.—CONTINENTAL ELEVATION AND DEPRESSION.

6. Before leaving this summary of causes, it is necessary to make a few general statements respecting elevation and depression. The first and most important is that, from the great Pliocene elevation onward, subsidence and re-elevation were always in progress. At each stage of these there must have been corresponding geographical conditions and varying facilities for distribution of travelled detritus. In regard, therefore, to the causes of any particular deposit, one of the most important

questions is, at what stage of elevation or depression was it produced. The second is that we must not infer that the elevations and depressions were necessarily uniform locally, but that they were different in amount in different places, and that elevation of mountainous regions often coincided with depression of plains, and *vice versâ*. The observations of Upham and Spencer as to what has been termed "warping" illustrate this, and it has been finally established by the work of Dr. G. M. Dawson on the Cordilleran glacier of the west already noticed.

It is not my purpose here to discuss the causes of the elevations and depressions of the continents in the later tertiary time. The attempt to account primarily for depression or elevation of the land or the ocean level by the accumulation of ice on the land is futile, since that accumulation itself must have depended largely on the changes of elevation and of consequent distribution of land and water. Primarily, the great elevation of the land must have been caused by the slow depression of the ocean bed in the intervals between those local foldings of the crust which result from and relieve such depression. That some later or secondary portion of the local differential depression of mountain regions may have been caused by the great weight of ice heaped on them is probable; but it is evident that this effect is quite inconsistent with the idea of wide-spread continental glaciers.

Students of glacial phenomena are no doubt right in directing attention to the great sensitiveness of the crust of the earth to pressure. The phenomena of river deltas and of such great thicknesses of sediment as those of the coal-formation, shows, as I have elsewhere often argued, that every foot of sediment placed on any part of the earth's crust must produce a corresponding depression either slow and gradual or by paroxysms, as the weight increases beyond the limit of the rigidity of the outer crust. Hence, a great weight of ice placed on mountains or high table-lands must tend to depress them relatively to the plains and sea beds, and the lateral pressure on the under crust may co-operate in raising the latter. Such movements, however, though important, must ever constitute a subsequent and incidental effect of glacial accumulations proceeding from other causes. In this connection it must also be observed that different portions of the crust must be of unequal thickness and hardness, and supported on material of different degrees of mobility; and further, that there are many fractures in the crust presenting lines of weakness. These differences must materially affect the results of pressure in different localities.

Movements of the kind above referred to have not ceased. Certain regions have in very recent times been, and are still being, weighed down by superficial accumulations, or are being buoyed up by the removal of matter by denudation or by the lateral pressure under them of the subterranean forces of the earth, while locally such effects are here and there being relieved by igneous eruptions. This is, however, a subject too large to be treated of here.

III.—Climatal Conditions.

We have now to consider the causes which could have led to such climatal conditions as those to which we have referred; and here, however unreasonable this may appear to some, I am disposed to content myself with the geographical changes long ago insisted on by Sir C. Lyell. There is the more reason to do this, since the facts established show that great geographical changes actually

occurred in the Pleistocene age, and that we have not
now to account for anything so extreme as a polar ice-cap
or a continental glacier. I have already directed attention
in connection with this part of the subject to the views
which I expressed many years ago, and to which I still in
the main adhere. I do not ask any geologists, more
especially those still affected with the superstitions of
continental glaciers and ice-caps, to accept these causes
as sufficient to account for the climatic changes evidenced
in geological time; but I must ask that they should fully
exhaust the influence of known changes of distribution of
land and water, and differential elevation and depression
of continental masses, before invoking other causes,
whether of cold or heat. I must also insist on their
admitting, at least as primary conditions in glaciation, not
merely cold but heat, and not merely elevation but
depression of land. In other words, there must be
evaporation as well as condensation, and the former
depends on the application of heat to water-surfaces
adjacent to those of precipitation.* On the other hand,
evaporation being provided, there must, in order to
establish a breeding-ground for glaciers, and to permit
their existence, be a low mean temperature and high land
capable of affording a condensing surface. The statements

---

* To American geologists I would recommend a course of reading
in " Whitney's Climatic Changes," though I do not agree with the
author in all his conclusions. In a paper in the Proceedings of the
Boston Society of Natural History (1890), Upham and Everett fully
admit the geographical causes of the glacial cold, and also the
existence of two periods of boulder distribution, separated by an interglacial period, though they do not appear to see the bearing of these
and other admissions on the validity of the theory of continental
glaciation. See also an important paper by Upham in the *American
Geologist*, December, 1890.

already made in my first chapter are, I think, sufficient to illustrate these views, and I may therefore here merely introduce a few remarks respecting variations of climate in the glacial age, referring to the map of the North American continent in the Pleistocene period at p. 77.

In the early glacial period, if we judge from the great accumulation of snow on the Cordillera of the west and the Laurentian highlands, the temperature must have been low. Similar evidence is afforded by the few species of shells found in the boulder-clay, which are of species now occurring in the cold waters of the Arctic regions loaded all summer with ice.* It would seem that to reduce the mean temperature of the sea to this extent it would be necessary that geographical changes should occur which would direct most of the warm equatorial water both from the North Atlantic and the North Pacific.

In the time of the lower Leda clay the temperature of the sea seems scarcely to have increased; but in the upper Leda clay we have a marine fauna identical with that of the colder waters of the present gulf and river St. Lawrence. One can to-day dredge in a living state off Metis in the river St. Lawrence all the species found in the upper Leda clay of the neighbouring coast. In like manner the vegetable remains of the upper Leda clay and its equivalent in the west are not arctic but boreal plants, and we should have to go near to the arctic circle, then as now, to find the true arctic flora. These facts, while they imply a mean temperature somewhat lower than that of the present day, show that the climate of the mid-Pleistocene was not an arctic one. It may have

* See list of fossils, *infra*.

been a little lower in mean temperature, but less extreme than that of North America at the present day. It is farther to be observed that the Pleistocene marine fauna is a little less boreal in New England than in the St. Lawrence valley, and that further north in Hudson's bay and the arctic coasts, it is not very dissimilar from that of the St. Lawrence.

In the later glacial period, that of the Saxicava sand, the great size and wide dispersion of boulders indicates much heavy field-ice, and, consequently, a low temperature of the sea, while the existence of local glaciers on the high lands not submerged, also indicates a low temperature. To this corresponds the vast predominance of the species *Saxicava rugosa* in the lower part of the Saxicava sand. There would, in this period, seem to have been fluctuations in temperature, due, perhaps to elevations and depressions of land, so that while in some of the raised beaches the indications of ice-drift are not so extreme as at present, on other levels there are gigantic boulders, and some of these carried far. Thus the later Pleistocene was characterized at once by great variations in the elevation of the land and by corresponding vicissitudes of climate.

These few remarks will, I think, suffice on this subject, when taken in connection with the facts and principles stated beforehand in chapter first.

An interesting illustration of the effects of varying distribution of land and water, may be taken from that warm period already alluded to as intervening between the glacial and modern times, and coinciding with the second continental period of Lyell, as evidenced by the distribution of marine animals at present on the coasts of Nova Scotia and New England. This peculiarity of distribution attracted my attention, as a collector of

marine animals, at Pictou, on Northumberland strait, as
long ago as 1840, and is thus referred to in a later
address.*

If we draw a straight line from the northern end of
Cape Breton, through the Magdalen islands, to the mouth
of the bay des Chaleurs, we have to the southward an
extensive semi-circular bay, 200 miles in diameter, which
we may call the great *Acadian bay*, and on the north the
larger and deeper triangular area of the gulf of St.
Lawrence. This Acadian bay is a sort of gigantic warm-
water aquarium, sheltered, except in a few isolated banks,
which have been pointed out by Mr. Whiteaves, from the
cold waters of the gulf, and which the bather feels quite
warm in comparison with the frigid and often not very
limpid liquid with which we are fain to be content in the
lower St. Lawrence. It also affords to the more delicate
marine animals a more congenial habitat than they can
find in the bay of Fundy, or even on the coast of Maine,
unless in a few sheltered spots, some of which have been
explored by Prof. Verrill. It is true that in winter the
whole Acadian bay is encumbered with floating ice, partly
produced on its own shores and partly drifted from the
north; but, in summer, the action of the sun upon its
surface, the warm air flowing over it from the neighbouring land, and the ocean water brought in by the strait of
Canseau, rapidly raise its temperature, and it retains this
elevated temperature till late in autumn. Hence the
character of its fauna, which is indicated by the fact
that many species of molluscs, whose headquarters are
south of cape Cod, flourish and abound in its waters.
Among these are the common oyster, which is especially

* See Address, by the author, to Nat. Hist. Society of Montreal,
1874.

abundant on the coasts of Prince Edward island and northern New Brunswick, the Quahog or Wampum shell, the *Petricola pholadiformis*, which, along with *Zirfea crispata*, burrows everywhere in the soft sandstones and shales; the beautiful *Modiola plicatula*, forming dense mussel-banks in the sheltered coves and estuaries; *Cytherea (Callista) convexa*; *Cochlodesma leana* and *Cummingia tellinoides*; *Crepidula fornicata*, the slipper-limpet, and its variety *unguiformis*, swarming especially in the oyster beds; *Nassa obsoleta* and *Buccinum cinereum*, with many others of similar southern distribution. Nor is the fauna so very meagre as might be supposed. My own collections from Northumberland strait include about fifty species of mollusks, and some not possessed by me have been found by Mr. Whiteaves. Some of these, it is true, are northern forms, but the majority are of New England species.

The causes of this exceptional condition of things in the Acadian bay carry us far back in geological time. The area now constituting the gulf of St. Lawrence seems to have been exempt from the great movements of plication and elevation which produced the hilly and metamorphic ridges of the east coast of America. These all die out and disappear as they approach its southern shore. The tranquil and gradual passage from the lower to the upper Silurian ascertained by Billings in the rocks of Anticosti, and unique in North America, furnishes an excellent illustration of this. In the Carboniferous period the gulf of St. Lawrence was a sea area as now, but with wider limits, and at that time its southern part was much filled up with sandy and muddy detritus, and its margins were invaded by beds and dykes of trappean rocks. In the Triassic age the red sandstones of that period were

extensively deposited in the Acadian bay, and in part have
been raised out of the water in Prince Edward Island,
while the whole bay was shallowed and in part cut off
from the remainder of the gulf by the elevation of ridges
of lower Carboniferous rocks across its mouth.   In the
Post-pliocene period, that which immediately precedes our
own modern age, as I have elsewhere shown,* there was
great subsidence of this region, accompanied by a cold
climate, and boulders of Laurentian rocks were drifted
from Labrador and deposited on Prince Edward Island
and Nova Scotia, while the southern currents, flowing up
what is now the bay of Fundy, drifted stones from the
hills of New Brunswick to Prince Edward Island.   At
this time the Acadian bay enjoyed no exemption from the
general cold, for at Campbelltown, in Prince Edward
Island, and at Bathurst, in New Brunswick, we find in the
clays and gravels the northern shells generally characteristic of the Post-pliocene; though perhaps the lists given
by Mr. Matthew for St. John and by Mr. Paisley for the
vicinity of Bathurst, may be held to show some slight
mitigation of the arctic conditions as compared with the
typical deposits in the St. Lawrence valley.   Since that
time the land has gradually been raised out of the waters,
and with this elevation the southern or Acadian fauna
has crept northward and established itself around Prince
Edward Island, as the Acadian bay attained its present
form and conditions.   But how is it that this fauna is now
isolated, and that intervening colder waters separate it
from that of southern New England ?   Verrill regards this
colony of the Acadian bay as indicating a warmer climate
intervening between the cold Post-pliocene period and the

---

* Notes on Post-pliocene of Canada, *Canadian Naturalist*, 1872.

present, and he seems to think that this may either have
been coincident with a lower level of the land sufficient to
establish a shallow water channel, connecting the bay of
Fundy with the Gulf, or with a higher level raising many
of the banks on the coast of Nova Scotia out of water.
Geological facts, which I have illustrated in my Acadian
Geology, indicate the latter as the probable cause. We
know that the eastern coast of America has in modern
times been gradually subsiding. Further, the remarkable
submarine forests in the bay of Fundy show that within
a time not sufficient to produce the decay of pine wood this
depression has taken place to the extent of at least 40 feet,
and probably to 60 feet or more.* We have thus direct
geological evidence of a former higher condition of the
land, which may when at its maximum have greatly
exceeded that above indicated, since we cannot trace the
submarine forests as far below the sea level as they actually extend. The effect of such an elevation of the land
would be not only a general shallowing of the water in the
bay of Fundy and the Acadian bay, and an elevation of its
temperature both by this and by the greater amount of
neighbouring land, but, as Prof. Verrill well states, it would
also raise the banks off the Nova Scotia coast, and extending south from Newfoundland, so as to throw the arctic
current further from the shore and warm the water along
the coasts of Nova Scotia and northern New England. In
these circumstances the marine animals of southern New
England might readily extend themselves all around the
coasts of Nova Scotia and Cape Breton, and occupy the
Acadian bay. The modern subsidence of the land would
produce a relapse toward the glacial age, the arctic cur-

* Acadian Geology, p. 29.

rents would be allowed to cleave more closely to the coast, and the inhabitants of the Acadian bay would gradually become isolated, while the northern animals of Labrador would work their way southward.*

Various modern indications point to the same conclusions. Verrill has described little colonies of southern species still surviving on the coast of Maine. There are also dead shells of these species in mud banks, in places where they are now extinct. He also states that the remains in shell-heaps left by the Indians indicate that even within the period of their occupancy some of these species existed in places where they are not now found. Willis has catalogued some of these species from the deep bays and inlets on the Atlantic coast of Nova Scotia, and has shown that some of them still exist on the Sable island banks.†

Whiteaves finds in the Bradelle and Orphan bank littoral species remote from the present shores, and indicating a time when these banks were islands, which have been submerged by subsidence, aided no doubt by the action of the waves.

It would thus appear that the colonization of the Acadian bay with southern forms belongs to the modern period, but that it has already passed its culmination, and the recent subsidence of the coast has, no doubt, limited the range of these animals, and is probably still favouring the gradual inroads of the Arctic fauna from the north, which, should this subsidence go on, will creep slowly back to re-occupy the ground which it once held in the Pleistocene time.

---

* Since my address of 1874, Ganong has further illustrated this subject in the Transactions of the Natural History Society of New Brunswick, and of the Royal Society of Canada.

† Acadian Geology, p. 37.

Such peculiarities of distribution serve to show the effects of even comparatively small changes of level upon climate, and upon the distribution of life, and to confirm the same lesson of caution in our interpretation of local diversities of fossils, which geologists have been lately learning from the distribution of cold and warm currents in the Atlantic. Another lesson which they teach is the wonderful fixity of species. Continents rise and sink, climates change, islands are devoured by the sea or restored again from its depths; marine animals are locally exterminated, and are enabled in the course of long ages to regain their lost abodes; yet they remain ever the same, and even in their varietal forms perfectly resemble those remote ancestors which are separated from them by a vast lapse of ages and by many physical revolutions. This truth which I have already deduced from the Pleistocene fauna of the St. Lawrence valley, is equally taught by the molluscs of the Acadian bay, and by their Arctic relatives returning after long absence to claim their old homes.

Still another lesson may be learned here. It appears that our present climate is separated from that of the glacial age by one somewhat warmer, which was coincident with an elevated condition of the land. Applied to Europe, as it might easily be, this fact shows the futility of attempting to establish a later glacial period between the Pleistocene and the present, in the manner attempted, as I must think on the slenderest possible grounds, by Prof. Geikie in his late work "The Great Ice Age."

The grandeur of those physical changes which have occurred since the present marine animals came into being is well illustrated by some other facts to which our attention has been directed. Recent excavations in the Montreal mountain have enabled Prof. Kennedy and Prof. Adams to observe marine shells and gravel at a still higher level than that of the old beach above Cote des Neiges, which was so long ago described by Sir Wm. Logan and Sir Charles Lyell. The new positions are 534 to about 600 feet above the sea. Let us place this fact along with the discovery of the skeleton of a whale at an elevation of 420 feet, as far west as Smith's falls, in Ontario, and with that recorded by Prof. Bell in the report of the Geological Survey for 1870-71, of the occurrence of these same shells on the high lands north of lake Superior, at a height which, taking the average of his measurements, is 547 feet above the sea level. Let us further note the fact, that in the hills behind Murray bay and at Les Eboulements I have recorded the occurrence of these remains at the height of at least 600 feet. We have, then, before us the evidence of the submergence of a portion of the North American continent, at least 1,000 miles in length and 400 miles in breadth, to a depth of more than a hundred fathoms, and its re-elevation, without any appreciable change in molluscan life.

IV.—Date of the Glacial Period.

The question of the time that has elapsed since the glacial period is closely connected with that as to the causes of the climatal changes involved. If these last were astronomical, and dependent, as Croll * has ably argued, on the varying eccentricity of the elliptical orbit in which our earth moves, along with the gradual procession of the equinoxes on the equator, then the culmination

---

* "Climate and Time in their Geological Relations."

of the last cold period must have been at least 100,000
years ago, and a period of 80,000 years may have elapsed
since the ice age began to give way to the present condition of things. If, on the other hand, we suppose that the
climatal change depended on variations in the heat of
the sun, we have no measure of time, for if these occur to
the extent required we do not know their periods or if
these have any regularity. We can only infer from the
fixity of solar heat within very narrow limits in historical
times that any material change must have occurred very
long ago.

Lastly, if with Lyell we have recourse to changes of
elevation and depression leading to different amounts of
heating surface and different distribution of oceanic and
atmospheric currents on the earth itself, geologists may
assign less or more time to such changes according as they
prefer to regard them as the results of secular or cata-
clysmic changes. Thus if we adopt the astronomical
theory we are shut up to a very ancient date. If we can
explain the facts by merely geological changes the date
becomes uncertain.

I have in previous publications * on this subject argued
that the amount of denudation which has occurred since
the glacial period is very small, that animal and vegetable
life have remained unchanged since the ice age, and that
such facts as we can measure in river erosion and changes
related to this, indicate but a short time. We may here
look at the last of these and cite a few facts.

In the case of the falls of Niagara, we know that these
have cut the present gorge from lake Ontario back to the

---

* Notes on Pleistocene of Canada, 1872, and later papers in *Canad.
Record of Science.*

11

whirlpool, and have cleaned out an old channel above
this, and cut back the present face of the fall some
distance since the close of the glacial period, and the
careful observations of Dr. Spencer* have shown that
the existing relations of the Niagara escarpment and the
lakes were established antecedent to the time when the
present fall was established. Claypole† has also shown
that the terraces with fresh-water shells on the Niagara
river prove that the retaining ridge between lakes Erie
and Ontario was then as high as now. As I have elsewhere argued also, the thickness of the harder bed, the
Niagara limestone, which the river has to cut, has, owing
to the southerly dip of the rocks, been increasing as the
falls were cut back, and there is reason to believe that a
part of the gorge above the whirlpool was formed in pre-
glacial times, and has merely been cleaned out by the
modern river. There is also some reason to believe that
the amount of water in the fall may have been greater in
the early modern period than now.

What, then, is the rate of recession of this great
cataract, and how long has it been cutting its gorge ?
The rate of cutting has been variousiy estimated at from
one foot to three feet annually; but the actual measured
rate for the last forty-two years, as given on the authority
of Mr. R. S. Woodward, of the U. S. Geological Survey,
is 2·4 feet, or nearly two-and-a-half feet per year. This
will give, say 12,000 to 15,000 years for the time required,
and, making allowance for the deductions above stated,
we may confidently affirm that the great cataract began
its labour somewhere between seven and twelve thousand

---

* Iroquois Beach. Trans. R. S. Can., 1889.

† *American Naturalist*, Oct., 1886. Trans. Geol. Socy., 1888.

years ago, and this must have been at the close of the glacial period, whatever views we may take of the nature of that period.

The estimate derived from Niagara is confirmed by the ingenious and careful calculations of Winchell* respecting the recession of the falls of St. Anthony, on the Mississippi, and by those of Andrews,† on the lake margins of lake Michigan. The former gives a period of between 6,276 and 12,103 years, or an average of 8,859 years. The latter gives a period of from 5,290 to 7,490 years. Humphreys and Abbott deduce similar figures from the rate of deposit of the delta of the Mississippi. Prestwich has deduced similar conclusions for England from his careful and detailed observations of the later Pleistocene deposits in that country.‡ His estimate of the final disappearance of the ice-age is from 8,000 to 10,000 years, and no English geologist is of greater experience and authority in Pleistocene Geology.

It may be objected that all these data are very uncertain. This is true, but since these and a vast number of facts of similar character which might be cited from different parts of the world all point in one direction, their cumulative evidence becomes very strong: on the one hand in proof that the close of the glacial period is very recent, and on the other that it must have been caused by telluric changes, and these, geologically speaking, not of very great magnitude.

With reference to the connection of man with the Pleistocene ice age, the present tendency of the geological facts is toward the conclusion that man had his origin in

---

* Journal Geol. Society, Nov., 1878.
† Trans. Chicago Academy, Vol. II.
‡ Journal of Geol. Soc. of London, Aug., 1887.

the post-glacial continental period, and that he survived
the great depressions and fluctuations of land which closed
that period and destroyed so many land animals his contemporaries in early times.   Many observers, however (as
Capellini,Whitney, Harvey, Habenecht, etc.), have adduced
evidence more or less doubtful of the existence of man in
the " first continental period," that of the later Pliocene.
Perhaps the most convincing evidence of such antiquity yet
adduced is that by Dr. Mourlon, of the Geological Survey
of Belgium,* from which it would appear that worked flints
and broken bones of animals occur in deposits, the relations of which would indicate that they belong either to
the base of the Pleistocene or close of the Pliocene.
They are imbedded in sands derived from Eocene and
Pliocene beds, and supposed to have been *remanié* by wind
action.   With the modesty of a true man of science,
Mourlon presents his facts, and does not insist too strongly
on the important conclusion to which they seem to tend,
but he has certainly established the strongest case yet on
record for the existence of Tertiary man.   With this
should, however, be placed the facts adduced in a similar
sense by Prestwich in his paper on the worked flints of
Ightham.†

Should this be established, the curious result will follow
that man must have been the witness of two great continental subsidences, that of the early Pleistocene and the
early modern, the former of which, and perhaps the latter
also, must have been accompanied with a great access of
cold in the Northern Hemisphere.   It seems, however,
more likely that the facts will be found to admit of a
different explanation.

---

* Bull. de l'Academie Roy. de Belgique, 1889.
† Journal London Geological Society, May, 1889.

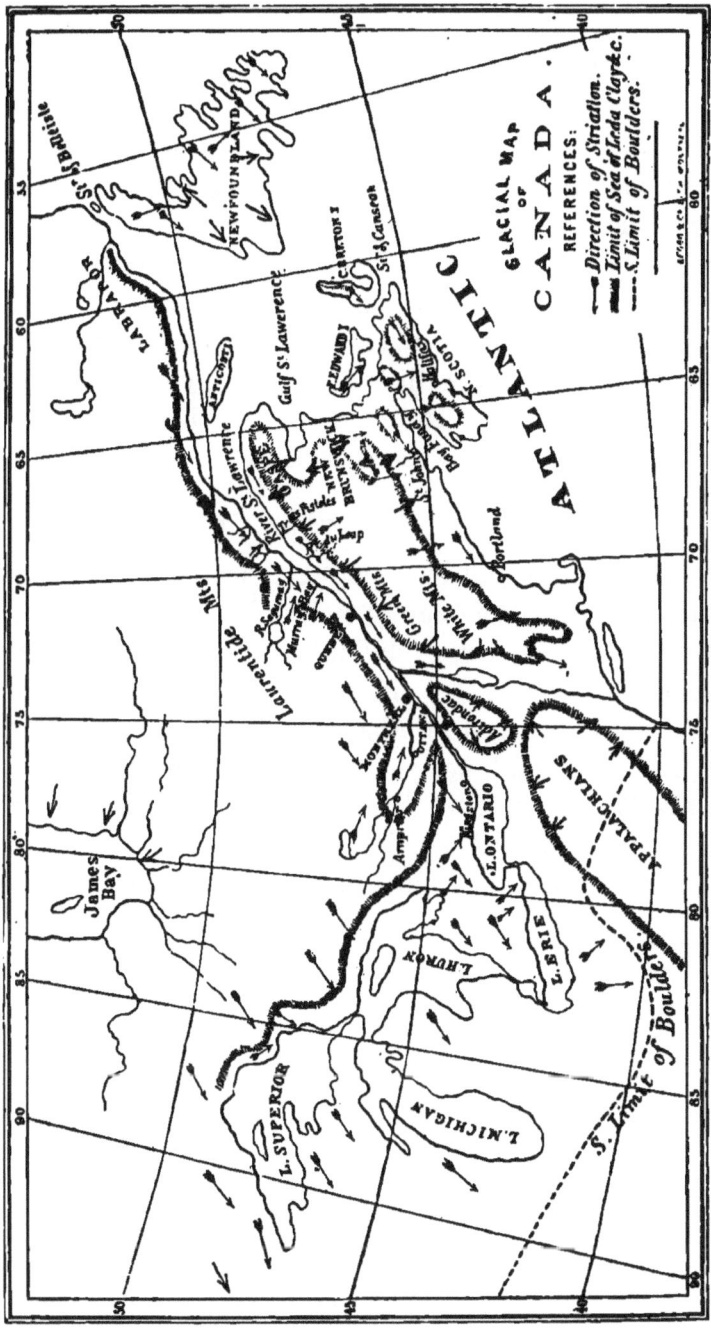

GLACIAL MAP
OF
CANADA.

REFERENCES:
Direction of Striation.
Limit of Sea of Leda Clay &c.
S. Limit of Boulders.

NOTE.—This map should have been entitled Glacial Map of *Eastern Canada.*

CHAPTER V.

SOME LOCAL DETAILS.

It will be impossible, in the space at my disposal, to embrace all the local details involved in my subject, and to give these would be tedious and unremunerative. For the greater part of them I must refer to reports and papers already in print, and some of which will be mentioned at the end of the chapter. I propose to notice only certain leading localities to which my own attention has been specially directed or which have important bearings on our general conclusions.

I.—*General Divisions of Canada.*

That northern half of North America included in the Dominion of Canada and Newfoundland may, for the purposes now in view, be divided geographically into six regions, characterized by distinctive physical features, and by distinct relations to the phenomena of the glacial age.

1. Newfoundland, not yet included in the Dominion of Canada, and separated from Labrador merely by the straits of Belle-isle, may be considered as an outlying part of the Laurentide range, and as an isolated centre of

ice distribution, proceeding in the early glacial period from its centre; and later as a series of reefs and islands in the arctic current.

2. The maritime provinces of Prince Edward Island, Nova Scotia and New Brunswick constitute a part of the Atlantic slope of North America, and not having any very high mountains, have had only minor centres of permanent ice, but have been traversed by powerful ice-laden currents from the north, and at certain periods of partial submergence, more especially in the later glacial age, have had boulders scattered over them from local sources in their own hills. Parts of northern New Brunswick have been invaded, in the more extreme glacial age, by local glaciers from the north-east extension of the Appalachian mountains.

3. The Canadian region proper, or that constituting the provinces of Quebec and Ontario, includes the wide valley of the gulf and river St. Lawrence, and part of the plateau of the great lakes. It curves to the south-west and north-west around the great salient angle of the Laurentian plateau, separating it from the basin of Hudson's bay and the Arctic sea, and has to the south-eastward the ridges of the Green and Appalachian mountains forming the breast-bone of the North American continent. In the earlier and more extreme glacial period its lower lands were submerged and received on their margins the ice discharged from the glaciers of the Laurentide hills, the Adirondacks and the Appalachians. The space between these now occupied by the St. Lawrence, was a great channel like Baffin's bay and Davis strait; but owing to its direction, much more intensely glaciated by floating ice borne on the Arctic current, which spread its burden over the submerged North

American plateau as far as the middle states of the Union, or to the lines of the modern Ohio and Missouri rivers.

4. The region of Manitoba and the North-west. This constitutes another, and now more elevated, plain, continuous with the former and with the great American plateau on the south, and extending north-west to the Arctic sea. It has the Laurentian axis on the north-east, and the Rocky mountains, the eastern ridges of the great Cordillera of the Pacific coast on the west. In the early Pleistocene, this great plain was at a lower level than at present, and the ice and *débris* from the Laurentide and Cordilleran glaciers, and more especially from the former, were distributed by water over its surface. In the mid-glacial age it was partially elevated and overspread with vegetation, but in the later glacial age it was much more extensively submerged and its waters covered with floating ice.

5. The great Cordilleran region of the west, embracing the Rocky mountains, the Gold and Selkirk ranges, the elevated interior plateau of British Columbia, and the coast ranges of the Pacific. In the early glacial period this region seems to have stood high out of the waters which extended to the east and west of it, and was covered with a great nevé, or snow-cap, sending off gigantic glaciers in all directions, but more especially to the south, north, and west. In the mid-glacial period it was greatly reduced in height, and, for the most part, denuded of ice, which, however, returned to it in diminished force in the second or later glacial age.

6. Lastly: Canada includes a portion of the Arctic basin north of the Laurentian ranges, and partly enclosed in the wide angle which they form northward. This, so far as known, was, throughout the glacial age, at a low level,

and with a climate little different from that which it at present possesses.* In the early glacial period, as well as in the great submergence of the later Pleistocene, its waters must have received contributions of ice, not only, as at present, from Greenland, but also from the northern parts of the Cordilleran and Laurentide glaciers, and there must have been immense accumulations of field-ice in the region of Hudson's bay and northward, which poured its superabundance around both ends of the Laurentides, and, in the times of greater submergence, probably forced its way through many gaps into the region of the great plains and the interior continental plateau south-west of the St. Lawrence valley and great lakes.

Bearing in mind these various local conditions, which result from the facts stated in previous chapters, we shall be prepared to appreciate the corroborative and otherwise interesting facts which appear in the following local details.

II.—*Newfoundland and Labrador.*

In the Journal of the Geological Society of London for February, 1871, is a communication from Staff-commander Kerr, R.N., in which he gives the directions of twenty-eight examples of grooved and scratched surfaces observed in the southern part of Newfoundland. The course of the majority of these is N.E. and S.W., ranging from N. 8° E. to N. 64° E. The remainder are N.W. and S.E., most of them with a predominating easterly direction. Boulders are mentioned, but no marine beds. The author refers

---

* This is proved by the transport of boulders to the north, by the temperate character of the flora to the south of it, and by the continued existence in it of the mammoth and its companions.

the glaciation to land ice shoving from the interior of the island, supposing certain submerged banks across the mouths of the bays to be terminal moraines. (See also the reports on the Geology of Newfoundland, by Murray and Howley.)

The latest information on the Pleistocene of Labrador is that given in a paper by Dr. Packard in the memoirs of the Boston Society of Natural History for 1867. The deposits are said to consist of boulders, Leda clay and sand, and raised beaches, which, on the authority of Prof. Hind, are stated to reach an elevation of 1,200 feet above the sea. The hills to a height of 2,500 feet are rounded as if by ice action. Some higher hills present a frost-shattered surface at their summits. No directions of striae are given, and they appear to be rare. Mr. Campbell, author of " Frost and Fire," mentions examples with course N. 45 °E. in the strait of Belle-isle. It is remarkable that true boulder-clay is rare in Labrador, though loose boulders are abundant in the valleys and on the inland table-land. Dr. Packard attributes the absence of boulder-clay to denudation. This may be the case, but it is to be observed that, on that view of the origin of boulder-clay which attributes it to ice-laden Arctic currents, there must always have been in the course of such currents areas of denudation as well as areas of deposition, and an elevated table-land like that of Labrador, in a high northern latitude, may well have been of the former character, or may have been a land area covered with snow and ice at the time of the deposition of the boulder-clay. In many respects, though less elevated, it resembles the aspect of the Cordillera region of the west as described by Dr. G. M. Dawson.

The Leda clay occurs in several places,   In 1860 I published a list of species collected by Capt. Orlebar; and Packard has greatly added to the number, giving a list which will be referred to farther on.   Dr. Packard very truly remarks that the fauna of the Labrador clays is very similar to that now found on the coast, and called by him the Syrtensian fauna.   In the latter we have a few southern forms, absent in the clay, but this is all.   Further, the Labrador pleistocene fauna is identical or nearly so with that of similar deposits in South Greenland, described by Möller and Rink.   Thus the climatal conditions of the Arctic current on the coast of Labrador seem to have in no respect differed in the Pleistocene from those which obtain at present.   The Leda clay with its characteristic fossils is found as high as 500 feet above the level of the sea.

Raised beaches and terraces, whether cut into sand and clay or the hard metamorphic rocks of the coast, are as common in Labrador as along the shores of the river St. Lawrence.   Their precise altitudes are not given, but they appear to be very numerous and rise to a great height above the sea.   One feature of some interest is their consisting in some places of large stones and boulders, evidencing very powerful action of coast ice and currents. Packard speaks of many of these beaches as glacier moraines modified by the sea.   From the descriptions of Prof. Hind,* it would also seem that there are traces of local glaciers in the river valleys, similar to those referred to above in the case of the Saguenay and the Murray river, and these might now be restored by a slight increase of cold or a moderate elevation of the land.

---

* Trans. Geol. Society, 1864.

III.—Anticosti.

On the island of Anticosti Messrs. Hyatt, Verrill and Shaler found *Saxicava arctica* in clay at an elevation of fifteen feet above the level of the sea.

The late Mr. Richardson of the Geological Survey, to whom we owe most of our knowledge of the geology of Anticosti, had previously noticed, in his Report for 1857, the occurrence of travelled boulders and of beds of clay, holding rounded fragments of limestone, and forming cliffs sixty to seventy feet high, but makes no mention of any pleistocene fossils. In 1885 the island was visited by Lt.-Col. Grant, of Hamilton, who made interesting collections, which he kindly presented to the Peter Redpath Museum of McGill University.*

The following are extracts from a letter of Col. Grant, referring to the localities of the fossils and the mode of their occurrence:

" The post-tertiary shells were first noticed in patches of blue clay in the south-west of Anticosti, in the bed of Becscia river, close to its mouth. When first seen, I thought it probable that they had been washed in by a high tide from the Gulf, but, on proceeding a short distance up stream, I found the clay and shells *in situ*, capped by a considerable thickness of drift, boulders, etc., in the river bank. The shells appeared to be unusually large. I collected a considerable number. Many got subsequently broken in rough weather.

" The pleistocene clay (Leda clay), occurs also in the bank and bed of Chaloupe river, and it is exposed along the cliff within a few miles west of the South-west point

---

* Notes on Pleistocene Fossils from Anticosti, by Lt.-Col. C. E. Grant and Sir W. Dawson, *Canadian Record of Science*, Vol. II., 1886.

158 THE ICE AGE IN CANADA.

lighthouse, and at several other points on the south shore.
On proceeding up Salmon river, north of Anticosti, at
about seven miles from the mouth, the high cliff on the
right bank is capped by a deposit of drift.

" Eight miles from the village of English bay (east), a
small stream from the top of the cliffs lays bare several
feet of blue clay, containing great numbers of very large
shells of Mya. The high tide reaches the base of the
clay and washes out numbers of specimens, as does the
brook adjacent. I was unable to examine the coast-line
except for a short distance. The cliffs, for some miles
beyond, from forty to seventy feet high, are crowned by
drift deposits. Where they slope, the boulders or rounded
pebbles from the top get mixed up with the clay below.
Fragments of shells are here numerous ; complete
specimens are few.

" The cliff to the west of Ellis or Gamache bay, called,
I think, ' Junction cliff,' by Richardson, is also crowned
by a drift deposit. I succeeded in reaching part of the
slope where some of the Leda clay from above had lodged.
I found it contained many specimens of *Saxicava rugosa*,
and a few of *Mya truncata*, the latter much smaller than
those at Becscia river and eight miles east of English bay.'
Glaciated or polished flags (chiefly Hudson river limestone) are not unusual in the drift of this part of the
island. Laurentian boulders were frequently remarked
in the river beds, some of considerable size also on the
land. There is one imbedded in the soil of a partly-
cleared farm near English bay.

" The island of Anticosti seems to be rising (the old
residents on various parts of the coast think the sea is
gradually retiring). I was assured by an inhabitant of
English bay, that the tops only of two large Laurentian

boulders, lying on the reef in front of the village, were visible at low water some twenty years ago; the base and many yards of the reef beyond are now exposed to view. A high ridge of shingle and sand in rear of the village represents the old beach. The bones of a whale were found on this beach. At Macdonald's cave, Mr. Macdonald, one of the oldest residents, informed me: 'This bay is filling up so fast that it will soon be dry land. I remember, when I first came here, there were about two or three feet of water where you now stand.' At Ellis bay, about twelve miles from English bay village, evidence also was obtained of the gradual elevation of the island."

The collection contains the following species, all of them previously known in the Pleistocene of other parts of Canada, and occurring as recent species in the colder waters of the gulf and river St. Lawrence :

Buccinum undatum, L., var. *labradoricum*. A small and somewhat short specimen, probably not fully grown.

B. glaciale, L. A decorticated shell, probably this species.

Trophon clathratum, L. (*T. scalariforme*, Gould). A well-developed specimen.

Natica affinis. One young shell.

Mya arenaria, L. Shells of moderate size, some of them distorted.

Mya truncata, L., var. *uddevalensis*. The short arctic variety, and one of them of unusually large size.

Macoma calcarea, Chem. Large specimens.

Macoma grœnlandica, L. One small valve.

Saxicava rugosa, L. Well-developed specimens and apparently common.

Astarte banksii, Leach. One valve.

Balanus crenatus, L.

Rhynchonella psittacea, L.

Col. Grant has also noted as occurring in the beds the following species, of which there are no specimens in the collection :

    Pecten islandicus.
    Mytilus edulis.
    Natica grœnlandica.
    Balanus hameri.

In sand and clay filling the interior of a Mya, which seems to have been entombed *in situ,* are many microscopic tests of foraminifera and valves of Cythere and Cytheridea. Among the former were the following species:

    Polystomella crispa.
    Nonionina scapha (and var. *labradorica*).
    Polymorphina lactea.
    Truncatulina lobata.
    Lagena sulcata.
    Entosolenia globosa.
    E. squamosa.
    Globigerina bulloides.

As usual in the Canadian Pleistocene, *Polystomella crispa* is much more abundant than the other species. *Nonionina scapha* comes next in this respect, and all the others are rare. The material also contains numerous spicules of siliceous sponges.

The above fossils may be regarded as characteristic of the Upper Leda clay and Saxicava sand, both of which members of the Pleistocene formation appear to be represented in Anticosti.

It would also appear that, as elsewhere in Canada, the Leda clay is overlaid by a second or newer boulder deposit connected with the Saxicava sand. To this it is probable that many of the travelled boulders of Lauren-

tian rocks belong, as they are found in this connection
not only along the whole south shore of the St. Lawrence,
but even in Prince Edward Island and in Nova Scotia. It
would be important to distinguish in Anticosti this upper
drift more particularly from the lower boulder-clay when
this may occur, and to observe any instances of glacial
striation.

With reference to the levels above the sea, it is to be
observed that along the shore of the St. Lawrence there
is usually a raised beach only a few feet above the level
of the sea, and on which shells and bones of whales
frequently occur, and a well-marked terrace, with beach
deposits and boulders, at a level of sixty or seventy feet
above the sea level, and this would appear to be the case
also in Anticosti.

Before proceeding up the St. Lawrence valley into
Canada proper, I may cross to the south side of the gulf
of St. Lawrence and notice the drift-deposits of Prince
Edward Island, Nova Scotia and New Brunswick, and
their connection with those of the state of Maine.

IV.—*Prince Edward Island.*

The Triassic and Permian rock formations of this
island consist almost entirely of red sandstones, and the
country is low and undulating, its highest eminences not
exceeding 400 feet. The prevalent Pleistocene deposit
is a boulder-clay, or in some places boulder loam, composed
of red sand and clay derived from the waste of the red
sandstones. This is filled with boulders of red sandstone
derived from the harder beds. They are more or less
rounded, often glaciated, with striae in the direction of
their longer axis, and sometimes polished in a remarkable
manner, when the softness and coarse character of the rock

12

are considered. This polishing must have been effected by rubbing with the sand and loam in which they are embedded. These boulders are not usually large, though some were seen as much as five feet in length. The boulders in this deposit are almost universally of the native rock, and must have been produced by the grinding of ice on the outcrops of the harder beds. In the eastern and middle portion of the island, only these native rocks were seen in the clay, with the exception of pebbles of quartzite which may have been derived from the Triassic conglomerates. At Campbellton, in the western part of the island, I observed a bed of boulder-clay filled with boulders of metamorphic rocks similar to those of the mainland of New Brunswick, to the southward of this locality.

Striae were seen only in one place on the north-eastern coast and at another on the south-western. In the former case their direction was nearly S.W. and N.E. In the latter it was S. 70° E.

No marine remains were observed in the boulder-clay; but at Campbellton, above the boulder-clay already mentioned, there is a limited area occupied with beds of stratified sand and gravel, at an elevation of about fifty feet above the sea, and in one of the beds there are shells of *Tellina Grœnlandica*.

On the surface of the country, more especially in the western part of the island, there are numerous travelled boulders, sometimes of considerable size. As these do not appear in situ in the boulder-clay, they may be supposed to belong to a second or newer boulder-drift similar to that which we shall find to be connected with the Saxicava sand in Canada. These boulders being of rocks foreign to Prince Edward Island, the question of their source becomes an interesting one. With reference to this, it may be stated in general terms that the majority are granite, syenite, diorite, felsite, porphyry, quartzite and coarse slates, all identical in mineral character with those which occur in the metamorphic districts of Nova Scotia and New Brunswick, at distances of from 50 to 200 miles to the south and south-west; though some of them may have been derived from Cape Breton on the east. It is further to be observed that these boulders are most abundant and the evidences of denudation of the Trias greatest in that part of the island which is opposite the deep break between the hills of Nova Scotia and New Brunswick, occupied by the bay of Fundy, Chignecto bay and the low country extending thence to baie Verte and Northumberland strait, an evidence that this boulder-drift was connected with currents of water passing up this depression from the south or south-west during, perhaps, the later part of the Pleistocene.*

Besides these boulders, however, there are others of a different character; such as gneiss, hornblende schist, anorthosite and Labradorite rock, which must have been derived from the Laurentian rocks of Labrador and Canada, distant 250 miles or more, to the northward. These Laurentian rocks are chiefly found on the north side of the island, as if at the time of their arrival the island formed a shoal, at the north side of which the ice carrying the boulders grounded and melted away. With reference to these boulders, it is to be observed that a depression of four or five hundred feet would open a clear passage for the arctic current entering the straits of

---

* I am informed that Mr. Chalmers has discovered striae on the rocks of this low isthmus, which would show the passage of heavy ice through it in Pleistocene times.

Belle-Isle, to the bay of Fundy; and that heavy ice carried by this current might, at the time of greatest depression, ground on Prince Edward Island, or be carried across it to the southward. If the Laurentian boulders came in this way, their source is probably 400 miles distant in the strait of Belle Isle. On the north shore of Prince Edward Island, except where occupied by sand dunes, the beach shows great numbers of pebbles and small boulders of Laurentian rocks. These are said by the inhabitants to be cast up by the sea or pushed up by the ice in spring. Whether they are now being drifted by ice direct from the Labrador coast, or are old drift being washed up from the bottom of the Gulf, which, north of the island, is very shallow, does not appear. They are all much rounded by the waves, differing in this respect from the majority of the boulders found inland.

The older boulder-clay of Prince Edward Island, with native boulders, must have been produced under circumstances of powerful ice-action, in which comparatively little transport of material from a distance occurred. If we attribute this to a glacier, then, as Prince Edward Island is merely a slightly raised portion of the bottom of the gulf of St. Lawrence, this can have been no other than a gigantic mass of ice filling the whole basin of the gulf, and without any slope to give it movement except toward the centre of this great though shallow depression. On the other hand, if we attribute the boulder-clay to floating ice, it must have been produced at a time when numerous heavy bergs were disengaged from what of Labrador was above water, and when this was too thoroughly enveloped in snow and ice to afford many travelled stones. Farther, that this boulder-clay is a sub-marine and not a sub-aerial deposit, seems to be rendered probable by the circumstance that many of the boulders of sandstone are so soft that they crumble immediately when exposed to the weather and frost.

The travelled boulders lying on the surface of the boulder-clay evidently belong to a later period, when the hills of Labrador and Nova Scotia were above water, though lower than at present, and were sufficiently bare to furnish large supplies of stones to coast-ice carried by the tidal currents sweeping up the coast, or by the arctic current from the north, and deposited on the surface of Prince Edward Island, then a shallow sand-bank. The sands with sea shells probably belonged to this period, or perhaps to the later part of it, when the land was gradually rising. Prince Edward Island thus appears to have received boulders from both sides of the gulf of St. Lawrence during the later Pleistocene period; but the greater number from the south side, perhaps because nearer to it. It thus furnishes a remarkable illustration of the transport of travelled stones at this period in different directions, and in the comparative absence of travelled stones in the lower boulder-clay, it furnishes a similar illustration of the homogeneous and untravelled character of that deposit, in circumstances where the theory of floating ice serves to account for it, at least as well as that of land-ice, and in my judgment, greatly better.

In these respects the Pleistocene of Prince Edward Island bears considerable resemblance to that of the lower grounds of Nova Scotia, where local material is prevalent in the lower part of the deposit, and travelled boulders from different directions occur in the upper bed.

V.—Nova Scotia and New Brunswick.

In these provinces the older geological structure is different from that in Prince Edward Island, the country consisting of Carboniferous and Triassic plains, with ranges of older hills, often metamorphic, and attaining elevations of 1,200 feet or more. It may, perhaps, be best in the first instance to present a summary of the phenomena, as I have given them in my Acadian Geology, and to add such additional facts and inferences as the present state of the subject may require.

The beds observed may be arranged as follows, in descending order:

1. Gravel and sand beds, and ancient gravel ridges and beaches, indicating the action of shallow water, and strong currents and waves. Travelled boulders occur in connection with these beds.

2. Stratified clay with shells, showing quiet deposition in deeper water.

3. Unstratified boulder-clay, indicating, probably, the united action of ice and water.

4. Peaty deposits, belonging to a land-surface preceding the deposit of the boulder-clay.

As the third of these formations is the most important and generally diffused in Nova Scotia and New Brunswick, we shall attend to it first, and notice the relation of the others to it.

The unstratified drift and boulder-clay, which occurs chiefly at the lower levels of the country, varies from a stiff clay to loose sand, and its composition and color generally depend upon those of the underlying and neighbouring rocks. Thus, over sandstone it is arenaceous, over shales argillaceous, and over conglomerate and hard slates pebbly or shingly. The greater number of the stones contained in the drift are usually, like the paste containing them, derived from the neighbouring rock formations. These untravelled fragments are often of large size, and are usually angular, except when they are of very soft material, or of rocks whose corners readily weather away. It is easy to observe, that on passing from a granite district to one composed of slate, or from slate to sandstone, the character of the loose stones changes accordingly. It is also a matter of familiar observation, that in proportion to the hardness or softness of the prevailing rocks, the quantity of these loose stones increases or diminishes. In some of the quartzite and granite districts of the Atlantic coast, the surface seems to be heaped with boulders with only a little soil in their interstices, and every little field, cleared with immense labour, is still half filled with huge white masses popularly known as " elephants." On the other hand, in the districts of soft sandstone and shale, one may travel some distance without seeing a boulder of considerable size. The boulders are, as usual, often glaciated or marked with ice-striae.

Though the more abundant fragments are untravelled, it by no means follows that they are undisturbed. They have been lifted from their original beds, heaped upon each other in every variety of position, and intermixed with sand and clay, in a manner which shows convincingly that the sorting action of running water has nothing to do with the matter; and this applies not only to stones of moderate size, but to masses of ten feet or more in diameter. In some of the carboniferous districts where the boulder-clay is thick, as for example, near Pictou harbour, it is as if a gigantic harrow had been dragged over the surface, tearing up the outcrops of the beds, and mingling their fragments in a rude and unsorted mass.

Besides the untravelled fragments, the drift always contains boulders derived from distant localities, to which in many cases we can trace them; and I may mention a few instances of this to show how extensive has been this transport of detritus.   In the low country of Cumberland there are few boulders, but of the few that appear some belong to the hard rocks of the Cobequid hills to the southward; others may have been derived from the somewhat similar hills of New Brunswick.   On the summits of the Cobequid hills and their northern slopes, we find angular fragments of the sandstones of the plain below, not only drifted from their original sites, but elevated several hundreds of feet above them.   To the southward and eastward of the Cobequids, throughout Colchester, Northern Hants, and Pictou, fragments from these hills, usually much rounded, are the most abundant travelled boulders, showing that there has been great driftage from this elevated tract.   Near the town of Pictou, where a thick bed of a sandy boulder-deposit occurs, this is filled with large masses of sandstone derived from the outcrops of the beds on higher ground to the north ; but with these are groups of travelled stones often in the lower part of the mass.   Near the steam ferry wharf, in the town of Pictou, I observed one such group, consisting of the following, all large boulders and lying close together—two of red syenite, six of gray granite, one of compact gray felsite, one of hard conglomerate, two of hard grit.   The two last were probably Lower Carboniferous, the others derived from the older crystalline rocks.   All may have been drifted by one berg or ice-floe from the flanks of the Cobequid range of hills, or from the similar hills to the east and south. In like manner, the long ridge of trap rocks, extending from cape Blomidon to Briar island, has sent off great quantities of boulders across the sandstone valley which bounds it on the south and up the slopes of the slate and granite hills to the southward of this valley. Well-characterized fragments of trap from Blomidon may be seen near the town of Windsor; and I have seen unmistakable fragments of similar rock from Digby neck, on the Tusket river, thirty miles from their original position. On the other hand, numerous boulders of granite have been carried to the northward from the hills of Annapolis, and deposited on the slopes of the opposite trappean ridge; and some of them have been carried round its eastern end, and now lie on the shores of Londonderry and Onslow. So also, while immense numbers of boulders have been scattered over the south coast from the granite and quartz rock ridges immediately inland, many have drifted in the opposite direction, and may be found scattered over the counties of Antigonish, Pictou and

Stratified gravel on boulder-clay, Merigonish, N.S.

Colchester. A few boulders, apparently of Laurentian rock from Labrador, occur on the north coast of Nova Scotia, and Dr. Honeyman has recorded similar boulders near Halifax on the Atlantic coast. These facts show that the transport of travelled blocks, though it may here as in other parts of America have been principally from the northward, has by no means been exclusively so; boulders having been carried in various directions, and more especially from the more elevated and rocky districts to the lower grounds in their vicinity. Professor Hind has shown the existence of a similar relation between the boulders of New Brunswick and the hilly ranges of that country.

Such observations as I have been able to make in Nova Scotia and New Brunswick, and those of Hind, Matthews and others, show a general southerly and south-easterly direction of striation, with some local variations. The Reports of Mr. R. Chalmers of the Geological Survey of Canada have, however, contributed a large mass of new material,* and have gone far to enable us to distinguish the effects of local glacier action from those of sea-borne ice. It would appear, from these observations, that while local glaciers have been shed in different directions, even from the comparatively low mountains of the maritime provinces, a large and even dominant influence has been exercised by marine agencies. The tables of striation in Mr. Chalmers' Report of 1888–9 are especially worthy of study in this respect. His general results for southern New Brunswick are thus stated: † "Co-ordinating the data at hand respecting the glaciation of the region, it would seem that the theory of local glaciers on the higher grounds and ice-bergs or floating ice upon the lower, during the post-tertiary submergence of these, is sufficient to account for all the facts coming under observation."

---

* Report Geol. Survey of Canada, 1885 and following years.

† Report of 1889, Ottawa, 1890. See also papers by Mr. Chalmers in *Canadian Record of Science*, and Trans. Royal Society of Canada, 1886, p. 139.

This general statement appears to me to apply throughout the maritime provinces, though there are many local complexities, owing to the peculiar orographical and geographical features of the region.

The following notes relate to a few special features referred to in my previously published papers, and to the occurrence of marine fossils in the maritime provinces.

The travelled and untravelled boulders are usually intermixed in the drift. In some instances, however, the former appear to be most numerous near the surface of the mass, and their horizontal distribution is also very irregular. In examining coast sections of the drift we may find for some distance a great abundance of angular blocks, with few travelled boulders, or both varieties are equally intermixed, or travelled boulders prevail; and we may often observe particular kinds of these last grouped together, as, for instance, a number of blocks of granite, greenstone, syenite, etc., near each other, as if they had been removed from their original beds and all deposited together at one operation. On the surface of the country where the woods have been removed, this arrangement is sometimes equally evident; thus hundreds of granite boulders may be seen to cumber one limited spot, while in its neighbourhood they are comparatively rare. It is also well known to the farmers in the more rocky districts that many spots which appear to be covered with boulders have, when these are removed, a layer of soil comparatively free from stones beneath. These appearances may in some instances result from the action of currents of water, which have in spots carried off the sand or clay, leaving the boulders behind; but in many cases this is manifestly the original arrangement of the material, the superficial layer of boulders belonging to a more recent driftage than that of the underlying mass in which boulders are often much less abundant.

Boulders or travelled stones are often found in places where there is no other drift. For example, on bare granite hills, about 500 feet in height, near St. Mary's river, there are large angular blocks of quartzite, derived from the ridges of that material, which abound in the district, but which are separated from the hills on which the fragments lie by deep valleys.

In Nova Scotia, beds with marine shells have been found by Mr. Matthew at Horton bluff, but not elsewhere, though the boulder-clay is often covered with beds of stratified sand and gravel. The only evidence of land life, in the boulder period, or immediately before it, that I have noticed, is a hardened peaty bed which appears under the boulder-clay on the north-west arm of the river of Inhabitants in Cape Breton. It rests upon gray clay similar to that which underlies peat bogs, and is overlaid by nearly twenty feet of boulder-clay. Pressure has rendered it nearly as hard as coal, though it is somewhat tougher and more earthy in appearance. It has a shining streak, burns with considerable flame, and approaches in its characters to the brown coals or more imperfect varieties of bituminous coal. It contains many small roots and branches, apparently of a taxine tree, with *débris* of swamp plants. The vegetable matter composing this bed must have flourished before the drift was spread over the surface.

In New Brunswick, stratified clays holding marine shells have been found overlying the boulder-clay, or in connection with it, especially in the southern part of the

* Geol. Survey of Canada, 1889 and previous years.

province, where deposits of this kind occur similar to those found in Canada and in Maine, though apparently on a smaller scale. These deposits, as they occur near St. John, consist of gray and reddish clays, holding fossils which indicate moderately deep water, and are, as to species, identical with those occurring in similar deposits in Canada and in Maine. They would indicate a somewhat lower temperature than that of the waters of the bay of Fundy at present, or about that of the northern part of the gulf of St. Lawrence.

In Bailey's Report on the Geology of Southern New Brunswick, Professor Hartt has given a list of the fossils of these beds, as seen at Lawlor's lake, Duck cove and St. John, which I re-published with some additions in Acadian Geology.

These New Brunswick beds are strictly continuous with, and equivalent to those which extend along the coast of New England, and thence ascend into the valley of lake Champlain, while on the other side they may be considered as perfectly representing in character and fossils the Leda clay of Eastern Canada. They are remarkably like both in mineral character and fossils to the Clyde beds of Scotland, which are probably their equivalents. The points of resemblance of the Leda clay of the coast of Maine, and that of the St. Lawrence, and Labrador, were noticed by me in my paper of 1860, already referred to, and have been more fully brought out by Dr. Packard, who describes the Leda clay as it occurs at several localities from Eastport to cape Cod. Along this whole coast it retains its Labradoric or gulf of St. Lawrence aspect, though with the introduction of some more southern species, and the gradual failure of some more arctic forms. South of cape Cod, as in the

modern sea, the Pleistocene beds assume a much more
southern aspect in their fossils, the boreal forms altogether
disappearing. For a very full exhibition of these facts, I
may refer to Dr. Packard's paper.

The stratified sand and gravel of Nova Scotia rests
upon and is newer than the boulder-clay, and is also
newer than the stratified marine clays above referred to.
Its age is probably that of the Saxicava sand of the St.
Lawrence valley. The former relation may often be seen
in coast sections or river banks, and occasionally in road
cuttings. I observed some years ago an instructive
illustration of this fact in a bank on the shore a little to
the eastward of Merigomish harbour, At this place the
lower part of the bank consists of clay and sand with
angular stones, principally sandstones. Upon this rests a
bed of fine sand and small rounded gravel with layers of
coarser pebbles. The gravel is separated from the drift
below by a layer of the same sort of angular stones that
appear in the drift, showing that the currents which
deposited the upper bed have washed away some of the
finer portions of the drift before the sand and gravel were
thrown down. In this section, as well as in most others
that I have examined, the lower part of the stratified
gravel is finer than the upper part, and contains more sand.

In some cases we can trace the pebbles of the gravels
to ancient conglomerate rocks which have furnished them
by their decay; but in other instances the pebbles may
have been rounded by the waters that deposited them in
their present place. In places, however, where old pebble
rocks do not occur, we sometimes find, instead of gravel,
beds of fine laminated sand. A very remarkable instance
of the connection of superficial gravels with ancient
pebble rocks occurs in the county of Pictou. In the coal

formation of this county there is a very thick bed of conglomerate, the outcrop of which, owing to its comparative hardness and great mass, forms a high ridge extending from the hill behind New Glasgow across the East and Middle rivers, and along the south of the West river, and then, crossing the West river, re-appears in Rogers' hill. The valleys of these three rivers have been cut through this bed, and the material thus removed has been heaped up in hillocks and beds of gravel, along the banks of the streams, on the side toward which the water now flows, which happens to be the north and north-east. Accordingly, along the course of the Albion Mines Railway and the lower parts of the Middle and West rivers, these gravel beds are everywhere exposed in the road-cuttings, and may in some places be seen to rest on the boulder-clay, showing that the cutting of these valleys was completed after the drift was produced. Similar instances of the connection of gravel with conglomerate occur near Antigonish, and on the sides of the Cobequid mountains, where some of the valleys have at their southern entrances immense tongues of gravel extending out into the plain, as if currents of enormous volume had swept through them from north to south.

The stratified gravels do not, like the older drift, form a continuous sheet spreading over the surface. They occur in mounds and long ridges, or eskers, sometimes extending for miles over the country. One of the most remarkable of these ridges is the " Boar's Back," which runs along the west side of the Hebert river in Cumberland. It is a narrow ridge, perhaps from ten to twenty feet in height, and cut across in several places by the channels of small brooks. The ground on either side appears low and flat. For eight miles it forms a natural road, rough, indeed, but

practicable with care to a carriage, the general direction
being nearly north and south. What its extent or course
may be beyond the points where the road enters on and
leaves it, I do not know; but it appears to extend from the
base of the Cobequid mountains to a ridge of sandstone that
crosses the lower part of the Hebert river. It consists of
gravel and sand, whether stratified or not I could not
ascertain, with a few large boulders. Another very singular ridge of this kind is that running along the west side
of Clyde river in Shelburne county. This ridge is higher
than that on Hebert river, but, like it, extends parallel to
the river, and forms a natural road, improved by art in
such a manner as to be a very tolerable highway. Along
a great part of its course it is separated from the river by
a low alluvial flat, and on the land side a swamp intervenes
between it and the higher ground. Shorter and more
interrupted ridges of this kind may also be seen in the
country northward and eastward of the town of Pictou.
In sections they are seen to be stratified, and they
generally occur on low or level tracts, and in places where,
if the country were submerged, the surf or marine currents
and tides might be expected to throw up ridges. The
presence of boulders shows that ice grounded on these
ridges, and it, probably by its pressure, in some instances,
modified their forms. These eskers, or "horse-backs,"
must not, however, be regarded as glacier moraines, to
which in structure they generally bear no resemblance.

Mr. Chalmers has in New Brunswick endeavoured, with
some success, to distinguish those that belong to river
valleys and glaciers from those that are marine.

The Rev. Mr. Paisley has published in the *Canadian
Naturalist* (1872) a list of shells obtained from a railway
cutting on the Tattagouche river, near Bathurst, in New

Brunswick. They were found in beds of Leda clay passing upwards into sand and gravel. At the Jacquet river, in the same district, the bones of a small cetacean have been found, and have been described by Dr. Gilpin and Dr. Honeyman.* They were referred by Dr. Gilpin to *Beluga Vermontana* of Thompson from the Pleistocene of Vermont. Similar bones have been found in the Leda clay of the St. Lawrence valley, and have been compared by the late Mr. Billings with the skeleton of the recent *B. catodon*, L., of the St. Lawrence, with which the so-called *B. Vermontana* is probably identical, as the specimens above referred to, and examined by Billings, certainly were.

Mr. Matthew has found *Tellina Grœnlandica* at Horton Bluff, in beds probably of the age of the Saxicava sand. Mr. Matthew has also published † a valuable synopsis of the fossils found up to 1876 in the Post-pliocene of New Brunswick, in which the number of species of mollusca is raised to more than thirty. He notes the important fact that the shells found on the coast of the baie de Chaleurs are of more northern type than those in the bay of Fundy, which conform more nearly to the assemblage found in these deposits on the New England coasts, so that the existing geographical regions were already to some extent established on the coast of North America in the period of the Upper Leda clay.

It is probably to the more modern part of the Pleistocene, if not to a more recent period following the elevation of the land, that the bones of the mastodon found in cape Breton, and described in "Acadian Geology," belong. To

---

* Trans. Nova Scotia Institute, Vol. III.
† *Canadian Naturalist*, Vol. VIII.

13

this later or post-glacial age also belong the boulder
pavements of lakes, the shore ridges, the oyster beds and
the sand dunes described in the same work and in the
" Supplement " to it (page 17).

VI.—Lower St. Lawrence—North Side.

Descriptions of the Pleistocene deposits of this region
are contained in several of my papers above cited, but I
shall here give a summary of these, with some corrections
and additional facts obtained within the past few years.

Saguenay River.—I have already, in part first, referred
to the glacial striation of this region, and perhaps no
better example could be found of those lateral valleys
along which ice seems to have been poured into the St.
Lawrence from the north.   The gorge of the Saguenay is
a narrow and deep cut, running nearly N.W. and S.E., or
at right angles to the course of the St. Lawrence, and of
the Laurentian ridges.   It extends inland more than
forty-five miles, and then divides into two branches, one
of which is occupied by the continuation of the river to
lake St. John, the other by Ha-Ha bay and a valley at its
head.   In the lower part of its course, as far as Ha-Ha
bay, this gorge is from 50 to 140 fathoms deep below the
level of the tide in the St. Lawrence, indicating an elevation of the land to that extent or more, at the time when
it was excavated.   In some places the cliffs on its banks
rise abruptly to 1,500 feet above the water level, so that
its extreme depth is nearly 2,400 feet, while its width
varies from about a mile to about one and a-half.   The
striated surfaces and the *roches moutonnées* seen in this
gorge and on the hills on its sides, to a height of at least
300 feet, shew that in the glacial period a powerful
stream of ice must have flowed down the gorge into the

St. Lawrence, though whether this was wholly a glacier or in part a fiord leading from one, like many of those in Greenland, does not certainly appear. Possibly, with different levels of the land, these conditions may have alternated. I cannot imagine anything more like what the Saguenay may have been, than the Franz Joseph fiord in east Greenland.*

The strikes of the gneiss on the opposite sides of the Saguenay indicate that it occupies a line of transverse fracture, constituting a weak portion of the Laurentian ridges, and this has evidently been smoothed and deepened by water and ice under conditions different from the present, in which it is probable that the channel is being gradually filled with mud. Its excavation must have taken place during a period of continental elevation in or after the Pliocene period, and previous to the deposition of the thick beds of marine clay (Leda clay) which appear near its mouth and in its tributaries, sometimes passing into boulder-clay below, and capped by sand and gravel. It is indeed not improbable that in the later Pleistocene it was in great part filled up with such deposits, which have been swept away in the course of the re-elevation of the land.

At Tadoussac, at the mouth of the Saguenay, where the underlying formation is the Laurentian gneiss, the Pleistocene beds attain to great thickness, but are of simple structure and only slightly fossiliferous. The principal part is a stratified sandy clay with few boulders, except in places near the ridges of Laurentian rocks, when it becomes filled with numerous rounded blocks and pebbles of gneiss. This forms high banks eastward of Tadoussac.

* Second German Expedition, 1870. See also a paper by Prof. Laflamme, Trans. R.S.C.

It contains a few shells of *Tellina Grœnlandica* and *Leda glacialis*, and a little inland, at Bergeron river, it also contains *Cardium Islandicum*, *Astarte elliptica*, and *Rhynchonella psittacea*. It resembles some of the beds seen on the south side of the river St. Lawrence, and has also much of the aspect of the Leda clay, as developed in the valley of the Ottawa. On this clay there rest in places thick beds of yellow sand and gravel.

At Tadoussac these deposits have been cut into a succession of terraces which are well seen near the hotel and old church. The lowest, near the shore, is about ten feet high; the second, on which the hotel stands, is forty feet; the third is 120 to 150 feet in height, and is uneven at top. The highest, which consists of sand and gravel, is about 250 feet in height. Above this, the country inland consists of bare Laurentian rocks. These terraces have been cut out of deposits, once more extensive, in the process of elevation of the land; and the present flats off the mouth of the Saguenay would form a similar terrace as wide as any of the others, if the country were to experience another elevatary movement. On the third terrace I observed a few large Laurentian boulders, and some pieces of red and gray shale of the Quebec group, indicating the action of coast-ice when this terrace was cut. On the highest terrace there were also a few boulders; and both terraces are capped with pebbly sand and well-rounded gravel, indicating the long-continued action of the waves at the levels which they represent.

Murray Bay, etc.—At Murray bay, Petit Mal bay, and Les Eboulements, as noticed above, the system of Pleistocene terraces is well developed. On the west side of Murray bay, the Cambro-Silurian rocks of White point, immediately within the pier, form a steep cliff, in the middle of which is a terraced step marking an ancient sea level. At the end nearest the pier the sea has again cut back to the old cliff, leaving merely a narrow shelf; but toward the inner side this shelf rapidly expands into the sandy flat along which the main road runs, and which is continuous with the lower plain extending all the way to the head of the bay. In this flat the upper portion of the Pleistocene deposit seems to consist principally of sand and gravel, resting on stony clay. In the former, which corresponds to the Saxicava sand of Montreal, I found only a few valves of *Tellina Granlandica*, which is still the most abundant shell on the modern beach. In the latter, corresponding to the Leda clay, which is best seen in some parts of the shore at low tide, I found a number of deep water shells of the following species, all of which, except *Spirorbis spirillum* and *Aphrodite Grœnlandica*, have been found in these deposits at Quebec and Montreal :

Fusus tornatus.
Trophon scalariforme.
Margarita helicina.
Cylichna occulta.
Pecten Islandicus.
Tellina calcarea.
Leda truncata.
Saxicava rugosa.
Aphrodite Grœnlandica.
Mytilus edulis.
Mya arenaria.
Balanus Hameri.
Spirorbis spirillum.
S. vitrea.
Serpula vermicularis.

These shells imply a higher beach than that of this lower flat, which is not more than thirty feet above the present sea level. Accordingly, above this are several higher terraces. (See Supra under " Terraces and Raised Beaches.") The second principal terrace, which forms a steep bank of clay some distance behind the main road, is 116 feet in height, and is of considerable breadth, and has on its front in some places an imperfect terrace at the height of 81 feet. It corresponds nearly in height with the shoulder over which the road from the pier passes. Upon it, in the rear of the property of Mr. Duberger, is a little stream which disappears underground, probably in a fissure of the underlying limestone, and returns to the surface only on the shore of the bay. Above this is a smaller and less distinct terrace, 139 feet high. Beyond this the ground rises in a steep slope, which in many places consists of calcareous beds, worn and abraded by the waves, but showing no distinct terrace; and the highest distinct shore mark which I observed is a narrow beach of rounded pebbles at the height of more than 300 feet; but above this there is a flat at the height of 448 feet. This beach appears to become a wide terrace further to the north, and also on the opposite side of the bay. It probably corresponds with the highest terrace observed by Sir W. E. Logan at bay St. Paul, and estimated by him at the height of 360 feet.

As already stated, three of the principal terraces at Murray bay correspond nearly with three of the principal shore levels at Montreal; and in various parts of Canada two principal lines of old sea beaches occur at about 100 to 150 feet, and 300 to 350 feet above the sea, though there are others at different levels.

In the Pleistocene period the valley of the Murray bay river has been filled, almost or quite to the level of the highest terrace, with an enormously thick mass of mud and boulders, washed from the land and deposited in the sea-bed during the long period of Pleistocene submergence. Through this mass the deep valley of the river has been cut, and the clay, deprived of support and resting on inclined surfaces, has slipped downward, forming strangely shaped slopes, and outlying masses, that have in some instances been moulded by the receding waves, or by the subsequent action of the weather, into conical mounds, so regular that it is difficult to convince many of the visitors to the bay that they are not artificial. Sir W. E Logan, in his report on the district, has, in my view, given the true explanation of these mounds, which may be seen in all stages of formation on the neighbouring hill-sides. Their effect to a geological eye is to give to this beautiful valley an unfinished aspect, as if the time elapsed since its elevation had not been sufficient to allow its slopes to attain to their fully rounded contour. This appearance is no doubt due to the enormous thickness of the deposit of Pleistocene mud, to the uneven surfaces of the underlying rock, and possibly also in part to the earthquake shocks which have visited this region.

At the mouth of the Murray Bay river, the boulder-clay rests directly on the striated rock-surfaces, and is a true till, filled with the Laurentian stones and boulders of the inland hills, though resting on Cambro-Silurian limestone. It is evidently marine, since it contains shells of *Leda glacialis;* and many of the stones are coated with Bryozoa and Spirorbis. It is also observable that on the N.E. sides of the limestone ridges the boulders are more numerous and larger. Above the boulder-clay may in some places be seen a stratified sandy clay, which further up the river attains to a great thickness. It contains *Saxicava rugosa*, *Tellina Grœn-landica* and *Tellina calcarea*, as well as *Leda glacialis*. The most recent deposit is a sand or gravel, often of consider-able thickness, and in some of the beds of gravel the pebbles are more completely rounded than those of the modern beach.

I have already stated my reasons for believing that the upper part of the valley of the Murray Bay river may have been the bed of a glacier flowing down from the inland hills toward the St. Lawrence. N.W. and S.E. striae attributable to this glacier were seen at an elevation of 800 feet, and the marine beds were traced up to almost the same height, above which, to a height of about 1,200 feet, loose boulders were observed and glaciated rock-surfaces, but no marine deposits. It is probable, there-fore, that at a time when the sea extended up to an elevation of 800 feet, the higher part of the valley may have been filled with land ice. Whether the bergs from this, drifting down toward the St. Lawrence, produced the N.W. striation observed at a lower level, or whether at a previous period, when the land was higher, the ice extended farther down, may admit of doubt. Certainly no land ice has extended to a lower level than about 800 feet since the deposition of the marine boulder and Leda clay.

Very large boulders occur in this vicinity. One observed on the beach on the east side of the bay, is an oval mass of lime felspar, thirty feet in circumference, lying like most other large boulders in this region, with its longer axis to the N.E.

Les Eboulements.—At this place the Laurentian hills rise to a great height near the shore, and the Pleistocene beds present the exceptional feature of resting on a soft and decomposed shale (Utica shale). This rock might indeed be mistaken for drift but for its stratification, and it must have been decomposed to a great depth by sub-aerial action and subsequently submerged and covered by the Pleistocene beds. Its preservation is the more remarkable that the clay overlying it contains very large Laurentian boulders, which must have been quietly deposited by floating ice. Only a few shells of *Tellina Grœnlandica* were observed in these clays.

The remarkable series of terraces seen at this place, and noticed in chapter second, rising to 900 feet in height, are all cut out of the Pleistocene beds and decomposed shale, and even the highest presents large boulders. In examining such terraces it is always necessary to distinguish between the clays out of which the terraces have been cut and the more modern deposits resting on the terraces. Both may contain fossils, but those of the original clay are in this region mostly of deeper water species than those in the overlying superficial beds.

I attribute the preservation of the thick beds of boulder-clay and the decomposed shale at Les Eboulements, to the fact that no transverse valley exists here, and that a point of high Laurentian land projects to the north-east, so as to shelter this place from forces acting in that direction. I have observed this appearance on the lee or south-west side of other projecting masses of hard rock, and as the decomposed shale must be a monument remaining from the Pliocene elevation of the land, it shows that no powerful eroding force had acted between that time and the period of the N.E. arctic ice-laden currents.

It is perhaps deserving of notice that the thick beds of soft material at Les Eboulements have been cut into many irregular forms by modern subaerial causes of denudation, and also by landslips; which last have been in part connected with the earthquake shocks with which this part of the coast has been visited more than any other district of Canada.

Above Les Eboulements, bay St. Paul presents features similar to those of Murray bay, and then the Laurentian land of cape Tourment comes boldly forward to the shore of the river. Above this the conditions are similar to those observed in the neighbourhood of Quebec.

VII.—Lower St. Lawrence.—South Side.

The Report of the Geological Survey of Canada (1863) includes all that is yet known of the Pleistocene formations at Gaspé, and thence upward to Trois Pistoles. According to this Report, the boulder-clay and overlying sands and gravels are extensively spread over the peninsula of Gaspé. On the Magdalen river they have been traced up to a height of 1,600 feet above the sea, though marine shells are not recorded at this great height. Terraces occur at various elevations, and in one of the lower at port Daniel, only fifteen feet above the sea, marine shells occur. On the coast westward of cape Rosier, terraces occur at many places, and of different heights, and marine shells have been found ninety feet above the sea. I have not had opportunities to examine these deposits to the eastward of the place next to be mentioned.

Trois Pistoles.—At this place one of the most complete and instructive sections of the Pleistocene in Canada has been exposed by the deep ravine of the river, and by the cuttings for the Intercolonial Railway. The most important terrace at the mouth of the Trois Pistoles river, that in which the railway cutting has been made, is about one hundred and fifty feet above the level of the sea, and is composed of clay, capped with sand and gravel. At no great distance inland, there rises a second terrace one hundred and sixty feet higher than the first, or about three hundred and ten feet above the sea. In some places the front of this terrace is cut into two or more. It consists of clay capped with sand and gravel, with some large stones and Laurentian boulders. Still farther inland is a third terrace, the height of which was estimated at four hundred to four hundred and fifty feet.

In the first mentioned of the above terraces, a very deep railway cutting has been made, exposing a thick bed of homogeneous clay of a purplish gray colour, and extremely tenacious. It contains few fossils; and these, as far as I could ascertain, exclusively *Leda glacialis.* It is, in short, a typical Leda clay, and its thickness in this lower terrace can scarcely be less than one hundred and twenty feet. As the inland terraces are probably also cut out of it, this may be less than half of its maximum depth. Under the Leda clay a typical boulder-clay had been exposed at one place in digging a mill sluice. It seemed to be about twenty feet thick, and rests on the smoothed edges of the shales of the Quebec group.

Though the Leda clay at the Trois Pistoles seems perfectly homogeneous, it shows indications of stratification, and holds a few large Laurentian boulders, which become more numerous in tracing it to the westward. A short distance westward of Trois Pistoles, it is seen to be overlaid by a boulder-deposit, in some places consisting of large loose boulders, in others approaching to the character of a true boulder-clay or associated with stratified sand and gravel. We thus have boulder-clay below, next Leda clay, and above this a second boulder-drift associated with the Saxicava sand, and apparently resting on the terraces cut out of the older clays. This is the arrangement which prevails throughout this part of Canada. It is modified by the greater or less relative thickness of the boulder-clay and Leda clay, by the irregular distribution of the overlying sands, and by the projection through it of ridges of the underlying rocks.

The section at Trois Pistoles may be represented as follows in descending order:

1. *Sand and Gravel*, capping the terraces cut in the previous deposits, and forming slight ridges or eskers in some of the lower levels. It contains on the lower terraces a few shells of *Leda* and *Tellina*. At the bottom of this deposit there are seen in places many large boulders of Laurentian and Lower Silurian rocks, resting on the Leda clay below.

2. *Leda Clay*, exposed in the railway cutting and seen also in the edge of the second terrace. Thickness one hundred and twenty feet or more. It holds a few large boulders and shells of *Leda glacialis*— the latter uninjured and with the valves united.

3. *Boulder-clay*, or hard gray till, with boulders and stones. Seen in a mill-sluice near the bridge, and estimated at twenty feet in thickness at this place, though apparently increasing in thickness farther to the westward.

4. *Shales* of Siluro-Cambrian age, seen in the bottom of the river near the bridge. They are smoothed over, but show no striae, though they have numerous structure lines which might readily be mistaken for ice-striae.

To the eastward of the mouth of Trois Pistoles river, the first terrace above mentioned is brought out to the shore by a projecting point of rock. In proceeding westward toward isle Verte, it recedes from the coast, leaving a flat of considerable breadth, which represents the lowest terrace seen on this part of the St. Lawrence, and is elevated only a few feet above the sea. This flat is in many places thickly strewn with large boulders, probably left when it was excavated out of the clay. In proceeding westward the first or railway terrace of Trois Pistoles, inland of the flat above mentioned, is seen to consist of boulder-clay, either in consequence of this part of the deposit thickening in this direction, or of the Leda clay passing into boulder-clay. It still, however, at isle Verte, contains a few shells of *Leda glacialis* in tough reddish clay holding boulders.

Rivière-du-Loup and Cacouna.—The country around Cacouna and Rivière-du-Loup rests on the shales, sandstones, and conglomerates of the Quebec and Potsdam groups of Sir W. E. Logan. As these rocks vary much in hardness, and are also highly inclined and much disturbed, the denudation to which they have been subjected has caused them to present a somewhat uneven surface. They form long ridges running nearly parallel to the coast, or north-east and south-west, with intervening longitudinal valleys excavated in the softer beds. One of these ridges forms the long reef off Cacouna, which is bare only at low tide; another, running close to the shore, supports the village of Cacouna; another forms the point which is terminated by the pier; a fourth rises into Mount Pilote; and a fifth stretches behind the town of Rivière-du-Loup.

The depressions between these ridges are occupied with Pleistocene deposits, not so regular and uniform in their arrangement as the corresponding beds in the great plains higher up the St. Lawrence, but still presenting a more or less definite order of succession. The oldest member of the deposit is a tough boulder-clay, its cement formed of gray or reddish mud derived from the waste of the shales of the Quebec group, and the stones and boulders with which it is filled partly derived from the harder members of that group, and partly from the Laurentian hills on the opposite or northern side of the river, here more than twenty miles distant. The thickness of this boulder-clay is, no doubt, very variable, but does not appear to be so great as farther to the eastward.

Above the boulder-clay is a tough clay with fewer stones, and above this a more sandy boulder-clay, containing numerous boulders, overlaid by several feet of stratified sandy clay without boulders; while on the sides of the ridges, and at some places near the present shore, there are beds and terraces of sand and gravel, constituting old shingle beaches apparently much more recent than the other deposits.

All these deposits are more or less fossiliferous. The lower boulder-clay contains large and fine specimens of *Leda glacialis* and other deep-water and mud-dwelling shells, with the valves attached. The upper clay is remarkably rich in shells of numerous species; and its stones are covered with Polyzoa and great Acorn-shells (*Balanus Hameri*), sometimes two inches in diameter and three inches high. The stratified gravel holds a few littoral and sub-littoral shells, which also occur in some places in the more recent gravel. On the surface of some of the terraces are considerable deposits of large shells of *Mya truncata*, but these are modern, and are the "kitchen-middens" of the Indians, who in former times encamped here.

Numbers of Pleistocene shells may be picked up along the shores of the two little bays between Cacouna and Rivière-du-Loup; but I found the most prolific locality to be on the banks of a little stream called the Petite Rivière-du-Loup, which runs between the ridge behind Cacouna and that of Mount Pilote, and empties into the bay between Rivière-du-Loup and the pier. In these localities I collected and noticed in my paper on this place * more than eighty species, about thirty-six of them not previously published as occurring in the Pleistocene of Canada.

We have thus at Rivière-du-Loup indubitable evidence of a marine boulder-clay, and this underlies the representative of the Leda clay, and rests immediately on striated rock surfaces, the striae running north-east and south-west.

The Cacouna boulder-clay is a somewhat deep-water deposit. Its most abundant shells are *Leda glacialis, Nucula tenuis,* and *Tellina proxima,* and these are imbedded in the clay with the valves closed, and in as perfect condition as if the animals still inhabited them. At the time when they lived, the Cacouna ridges must have been reefs in a deep sea. Even Mount Pilote has huge Laurentian boulders high up on its sides, in evidence of this. The shales of the Quebec group were being wasted by the waves and currents; and while there is evidence that much of the fine mud worn from them was drifted far to the south-west to form the clays of the Canadian plains, other portions were deposited between

---

* *Canadian Naturalist,* April, 1865.

the ridges, along with boulders dropped from the ice which drifted from the Laurentian shore to the north. The process was slow and quiet; so much so, that in its later stages many of the boulders became encrusted with the calcareous cells of marine animals before they became buried in the clay. No other explanation can, I believe, be given of this deposit; and it presents a clear and convincing illustration, applicable to wide areas in Eastern America, of the mode of deposit of the boulder-clay.

A similar process, though, probably, on a much smaller scale, is now going on in the Gulf. Admiral Bayfield has well illustrated the fact that the ice now raises, and drops in new places, multitudes of boulders, and I have noticed the frequent occurrence of this at present on the coast of Nova Scotia. At Cacouna itself, there is, on some parts of the shore, a band of large Laurentian boulders between half tide and low-water mark, which are moved more or less by the ice every winter, so that the tracks cleared by the people for launching their boats and building their fishing-wears, are in a few years filled up. Wherever such boulders are dropped on banks of clay in process of accumulation, a species of boulder-clay, similar to that now seen on the land, must result. At present such materials are deposited under the influence of tidal currents, running alternately in opposite directions; but in the older boulder-clay period, the current was probably a steady one from the north-east, and comparatively little affected by the tides.

The boulder-clay of Cacouna and Rivière-du-Loup, being at a lower level and nearer the coast than that found higher up the St. Lawrence valley, is probably newer. It may have been deposited after the beds of boulder-clay at Montreal had emerged. That it is thus more recent, is further shown by its shells, which are, on the whole, a more modern assemblage than those of the Leda clay of Montreal. In fossils, as well as in elevation, these beds more nearly resemble those on the coast of Maine. It would thus appear that the boulder-clay is not a continuous sheet or stratum, but that its different portions were formed at different times, during the submergence and elevation of the country; and it must have been during the latter process that the greater part of the deposits now under consideration were formed.

The assemblage of shells at Rivière-du-Loup is, in almost every particular, that of the modern gulf of St. Lawrence, more especially on its northern coast. The principal difference is the prevalence of *Leda arctica* in the lower part of the deposit. This shell, still living in Arctic America, has not yet occurred in the gulf of St. Lawrence, but is distributed throughout the lower part of the Pleistocene deposits in the whole of Lower Canada and New England, and appears in great numbers at Rivière-du-Loup, not only in the ordinary form, but in the shortened and depauperated varieties which have been named by Reeve *L. siliqua* and *L. sulcifera*.

Of *Astarte Laurentiana*, supposed to be extinct, and which occurs so abundantly in the Pleistocene at Montreal, few specimens were found, and its place is supplied by an allied but apparently distinct species, to be noticed in the sequel, which is still abundant at Gaspé and Labrador, and on the coast of Nova Scotia.

It must be observed that though the clays at Rivière-du-Loup are more recent than those of Montreal, they are still of considerable antiquity. They must have been deposited in water perhaps fifty fathoms deep, and the bottom must have been raised from that depth to its

14

present level; and in the meantime the high cliffs now
fronting the coast must have been cut out of the rocks of
the Quebec group.

The order of succession of beds, as seen in the banks of
the Little Rivière-du-Loup, may be stated as follows, in
descending order:

1. Large loose boulders, mostly of Laurentian rocks, seen
   in the tops of ridges of rock and gravel. One angular mass of Quebec group conglomerate was observed
   ninety feet in circumference and ten to fifteen feet
   high. Near it was a rounded boulder of Anorthosite
   from the Laurentian, 13 feet long.

2. Stratified sand and gravel resting on the sides of the
   ridges of rock projecting through the drift. Thickness variable.

3. Stratified sandy clay and sand with *Tellina Grœnlandica*
   and *Buccinum*. 10 feet.

4. Gray clay and stones. *Rhynconella psittacea*, and
   Terebratulina Spitzbergensis, &c. 1 foot or more.

5. Gray clay with large stones, often covered with Bryozoa
   and Acorn-shells. *Tellina calcarea* very abundant,
   also *Leda arctica*. 3 feet.

6. Tough, hard, reddish clay, with stones and boulders,
   passing downward into boulder-clay, and holding
   Leda arctica. 6 feet or more.

It was observable that the boulders were more abundant
on the south side of the ridges than on the north; and
between Rivière-du-Loup and Quebec there are numerous
small ridges and projecting masses of rock rising above
the clays, which generally show the action of ice on their
N.E. sides; while the large boulders lying on the fields are
seen to have their longer axes N.E. and S.W.

At the Petite Rivière-du-Loup the surface of the red clay (No. 6 above) was observed to have burrows of *Mya arenaria* with the shells (of a deep-water form) still within them.

I have already had occasion, in Chapter III., to notice the Pleistocene and modern deposits as seen at Little Metis, and may refer to that chapter for such details as are of interest.

VIII.—River St. Lawrence above Quebec, and Ottawa Valley.

Quebec and its Vicinity.—The deposits at Beauport, near Quebec, were described by Sir C. Lyell in the Geological Transactions for 1839; and a list of their fossils was given, and was compared with those of Montreal in my paper of 1859. As exposed at the Beauport Mills, the Pleistocene beds consist of a thick bed of boulder-clay, on which rests a thin layer of sand with *Rhynconella psittacea* and other deep-water shells. Over this is a thick bed of stratified sand and gravel filled with *Saxicava rugosa* and *Tellina*. Scattered Laurentian boulders here, as at Montreal and elsewhere, occur in the beds with the shells. In a brook near this place, and also in the rising ground behind Point Levis, the deep-water bed attains to greater thickness, but does not assume the aspect of a true Leda clay. Above Quebec, however, the clays assume more importance; and between that place and Montreal are spread over all the low country, often attaining a great thickness, and not unfrequently capped with Saxicava sand. At Cap à la Roche the officers of the Geological Survey have found a bed of stratified sand under the Leda clay. The Beauport deposit is evidently somewhat exceptional in its want of Leda clay, and this I suppose may have been owing to the powerful currents of water which have swept around Cape Diamond at the time of the elevation of the land out of the Pleistocene sea. The layer of sand at the surface of the boulder-clay is evidently here the representative of the Leda clay, and affords its characteristic fossils, while the stones projecting above the boulder-clay are crusted with Bryozoa and Acorn-shells. At St. Nicholas, there is a sandy boulder-clay, not unlike that of Rivière-du-Loup, which has afforded some very interesting fossils. It is stated in the Report of the Survey to be one hundred and eighty feet above the sea.

Montreal.—In the neighbourhood of Montreal very interesting exposures of the Pleistocene beds occur, and with the terraces on the Mountain have been described in my papers of 1857 and 1859. I may here merely condense the leading facts, adding those more recently obtained.

An interesting section of the deposits is that obtained at Logan's Farm, which may be thus stated in descending order :

	ft.	in.
Soil and sand	1	9
Tough reddish clay	0	0½
Gray sand, a few specimens of *Saxicava rugosa, Mytilus edulis, Tellina Grœnlandica,* and *Mya arenaria,* the valves generally united	0	8
Tough reddish clay, a few shells of *Astarte Laurentiana,* and *Leda glacialis*	1	1
Gray sand, containing detached valves of *Saxicava rugosa, Mya truncata,* and *Tellina Grœnlandica :* also *Trichotropis borealis,* and *Balanus crenatus ;* the shells, in three thin layers	0	8
Sand and clay, with a few shells, principally *Saxicava* in detached valves	1	3
Band of sandy clay, full of *Natica clausa, Trichotropis borealis, Fusus tornatus, Buccinum glaciale, Astarte Laurentiana, Balanus crenatus,* &c., &c., sponges and *Foraminifera.* Nearly all the rare and deep-sea shells of this locality occur in this band	0	3
Sand and clay, a few shells of *Astarte* and *Saxicava,* and remains of sea-weeds with *Lepralia* attached ; also *Foraminifera*	2	0
Stony clay (Boulder-clay). Depth unknown.		

In this section the greater part of the thickness corresponds to the Leda clay, which at this place is thinner and more fossiliferous than usual. Along the south-east side of the Mountain, and in the city of Montreal, the beds have been exposed in a great number of places, and are in the aggregate at least 100 feet thick, though the thickness is evidently very variable. The succession may be stated as follows :

1. *Saxicava Sand.*—Fine uniformly grained yellowish and gray silicious sand with occasional beds of gravel in some places, and a few large Laurentian boulders, *Saxicava, Mytilus,* &c., in the lower part. Thickness variable, in some places 10 feet or more.

2. *Leda Clay.*—Unctuous gray and reddish calcareous clay, which can be observed to be arranged in layers varying slightly in colour and texture. Some of these layers have sandy partings in which are usually Foraminifera and shells or fragments of shells. In the clay itself the only shells usually found are *Leda arctica* and a smooth deep-water form of *Tellina Grœnlandica;* but toward the surface of the clay, in places where it has not been denuded before the deposition of the overlying sand, there are many species of marine shells. A few large boulders are scattered through the Leda clay.

3. *Boulder-clay.*—Stiff gray stony clay, or till, with large boulders and many glaciated stones, often of the same Trenton rocks which occur on the flanks of the mountain. It is of great thickness, though it has been much denuded in places, and has not been observed to contain fossils. It is especially thick at the south and south-west sides of the Montreal mountain.

The Montreal mountain, like other isolated trappean hills in the great plain of the lower St. Lawrence, presents a steep, craggy front to the north-east, and a long slope or tail to the south-west; and in front of its north-east side is a bare, rocky plateau of great extent, and at a height of about 200 feet above the river. This plateau must have been produced by marine denudation of the solid mass of the mountain in the Pleistocene period, and proves an astonishing amount of this kind of erosive action in hard limestones interleaved with trap dykes, and which have been ground and polished with ice at the same time that the plateau was cut into the hill. By ice, also, must the *débris* produced by this enormous erosion have been removed, and piled along the more sheltered sides of the hill in the boulder-clay.

With regard to the crag-and-tail attitude of Montreal mountain, I have to observe that in large masses of this kind reaching to a considerable height, and rising above the Pleistocene sea, the north-east, or exposed, side has been cut into steep cliffs, but in smaller projections of the surface over which the ice could grind, the exposed side is smoothed, or "moutonnée," and the sheltered side is angular. A little reflection must show that this must be the necessary action of a sea burdened with heavy floating ice.

These facts have been well illustrated in the extensive limestone quarries lying on the plateau already referred to behind the city of Montreal, and north-east of the Montreal mountain. At this place the surface of the limestone has been polished and striated, the direction of the striae ranging from N. 50° E. to N. 70° E. Not only has the surface been intensely glaciated, but ledges of rock of great size have been lifted up and pushed to the S.W.

toward the mountain. On the glaciated surface lies boulder clay holding local and travelled boulders. Above this is clay and sand in layers, with numerous shells of many species, capped by sand and gravel filled with Saxicava. On the surface are here and there groups of Laurentian boulders. A few years ago all these appearances were well seen, but at present the quarrying operations have been carried toward the S. E. side of the ridge where the eroded surface of the limestone is covered only with a little soil. Farther opening to the eastward may again expose the glacial deposits. The appearances at this place are, I think, conclusive as to the action of floating ice drifting up the river valley from the N.E. The direction of striation and of movement of material and the marine character of the deposits all testify to this.

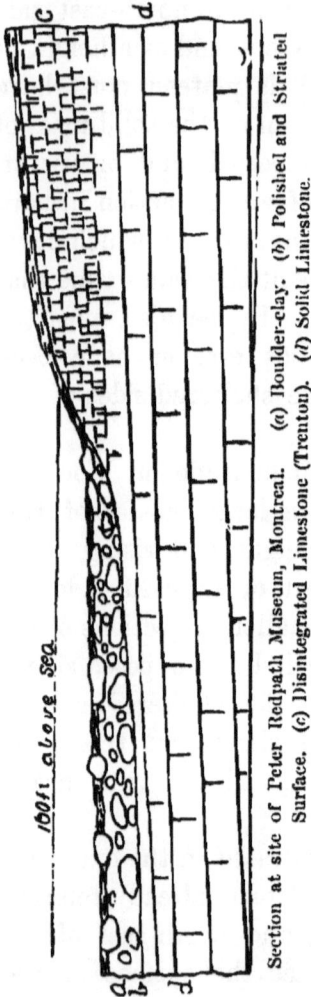

We have already seen that beaches holding littoral shells occur at Montreal mountain south-west of this limestone plateau up to a height of 470 feet, and at one place in the mountain toward its northern side there is a small plain, in the subsoil of which there occur shallow-water shells at an elevation of about 560 feet above the sea. Laurentian boulders, probably drifted on ice in the later glacial age, are found at a still higher level.

The site of the Peter Redpath Museum presents another interesting example of the special features of the drift-deposits on the south side of Montreal mountain. The first floor of the museum is 160 feet above the level of the sea, which is about the height to which intense glaciation and boulder-clay extend on the mountain,* the terraces above this level being of sand and gravel, and the limestone and trap of the mountain weathered and deeply decomposed and not covered with the boulder-clay. Thus the foundations of the building were excavated into a slope at the exact junction of the glaciated and non-glaciated surface. The excavation for the front of the building was made in tough boulder-clay, with large Laurentian and limestone boulders, and this rested on an intensely glaciated rock surface of limestone, with striae bearing S. 33° W. The rear of the building was cut into the same limestone, not glaciated, and decomposed to a depth of 20 feet or more into an earthy, crumbling mass, still showing the stratification and fossils of the formation.

There could not be a finer illustration of the "ice-foot" of the margin of the old Pleistocene sea; and any idea of glacier action was excluded by the directions of the striae, and by the absence of any lateral moraine.

The most strongly marked terraces on the Montreal mountain are at heights of 470, 440, 386, and 220 feet above the sea, but there are less important intermediate

---

* The heavy glaciation on the plateau north-east of the mountain extends up to about 180 feet.

terraces. The highest terrace holds littoral marine shells, which also occur on a little plateau at a height of 560 feet. On the highest of these, on the west side of the mountain, over Cote des Neiges village, there is a beach with marine shells, and on the summit of the mountain, at a height of about 750 feet, there are rounded surfaces, possibly polished by floating ice at the time of greatest depression, though no striation remains, and large Laurentian boulders, which must have been carried probably a hundred miles from the Laurentian regions to the N.E., and over the deep intervening valley of the St. Lawrence.*

I have already, in the earlier part of this section, noticed the striation on rock surfaces at Montreal, and may merely add that it is often very perfect, and must have been produced by a force acting up the St. Lawrence valley from the north-east, and planing all the spurs of the mountain on that side, while leaving the mountain itself as a bare and rugged unglaciated escarpment. In the streets of Montreal the true boulder-clay is often exposed in excavations, and is seen to contain great numbers of glaciated stones, most of which are of the hardened Lower Silurian shales and limestones of the base of the mountain; and, though no marine shells have been found, the sub-aquatic origin of the mass is evidenced by its gray unoxidised character, and by the fact that many of the striated stones at once fall to pieces when exposed to the frost, so that they cannot possibly have been glaciated by a sub-aerial glacier.

At the Glen brick-work, near Montreal, the Leda clay and underlying deposits have been excavated to a consid-

---

* Lyell ("Travels in North America," vol. 2, p. 140) very well describes the Pleistocene of the vicinity of Montreal.

erable depth, and present certain remarkable modifications. The section observed at this place is as follows :

	ft.	in.
1. Hard gray laminated clay, *Foraminifera* and *Leda*, in thin layers	7	0
2. Red layer, in two bands	0	6
3. Sandy clay	1	0
4. Gray and reddish clay	9	0
5. Hard buff sand, very fine and laminated	15	0
6. Sand with layers of tough clay, holding glaciated stones, and very irregularly disposed	4	0
7. Fine sand	1	0
8. Gray sand, with rounded pebbles, and laminated obscurely and diagonally	4	0
9. Fine laminated yellow sand	3	0
10. Gravel	0	4
11. Very irregular mass of laminated sand, with mud, gravel, stones and large boulders	12	0
	56	10

The whole of these deposits, except the Leda clay, are very irregularly bedded, and are apparently of a littoral character. They seem to show the action of ice in shallow water before the deposition of the Leda clay. The only way of avoiding this conclusion would be to suppose that the underlying beds are really of the age of the Saxicava sand, and that the Leda clay has been placed above them by slipping from a higher terrace; but I failed to see good evidence of this. A little farther west, at the gravel pits dug in the terrace for railway ballast, a deep section is exposed, showing at the top Saxicava sand, and below this a very thick bed of sandy clay with stones and boulders, constituting apparently a somewhat arenaceous and partially stratified equivalent of the boulder-clay. A little above this place, at the brick-works, the Saxicava sand is seen to rest on a highly fossiliferous Leda clay, which probably here intervenes between the two beds seen in contact nearer the edge of the terrace.

Ottawa River.—The Leda clay and Saxicava sand are well exposed on the banks of the Ottawa; and Green's creek, a little below Ottawa city, has become celebrated for the occurrence of hard calcareous nodules in the clay, containing not only the ordinary shells of this deposit, but also well-preserved skeletons of the Capelin (*Mallotus*), of the Lump-sucker (*Cyclopterus*), and of a species of stickle-back (*Gasterosteus*), of a *Cottus*, and of a species of seal. Some of these nodules also contain leaves of land plants and fragments of wood, and a fresh-water shell of the genus Lymnea has also been found.* At Packenham Mills, west of the Ottawa, the late Sheriff Dickson found several species of land and fresh-water shells associated with *Tellina Grœnlandica* and apparently in the Saxicava sand. These facts evidence the vicinity of the Laurentian shore, and indicate a climate only a little more rigorous than that of Central Canada at present. They were noticed in some detail in my paper of 1866 in *The Canadian Naturalist*.

Another illustration of the margin of the sea in this direction is afforded by the discovery of the bones of a whale at Smith's Falls, Ontario, in a bed of gravel, with a few marine shells, lying on the margin of the old Laurentian shore in this locality at a height of 420 feet above the level of the sea, an elevation not very different from that of one of the principal terraces with sea shells on Montreal mountain.

The marine deposits on the St. Lawrence are limited, as already stated, to the country east of Kingston; and the clays of the basin of the great lakes to the south-westward

* See notices of these fossils in Chapter V.

have, as yet, afforded no marine fossils.   Prof. Bell, of the
Geological Survey, has, however, found that two hundred
miles north of lake Superior the marine deposits reappear.

In the above local details, I have given merely the facts
of greatest importance, and may refer for many subordinate points to the papers catalogued in the introduction to this memoir, and to the reports of the Geological
Survey of Canada.

IX.—Western Districts.

In the Province of Ontario, west of the marine deposits,
which may be roughly stated to extend as far as Kingston,
the upper and lower drift are developed much in the same
manner as to the eastward, and contain many travelled
boulders from the Laurentian country to the north.   The
middle Pleistocene deposit, however, corresponding to the
Leda clay, and the greater part of which has been desig-
nated the Erie clay, is not only destitute of marine fossils,
but contains so little protoxide of iron that when burned it
does not assume a red colour, and it also contains fossil
plants, which will be noticed in the sequel, becoming thus
a "forest bed" or interglacial deposit.   The plants are of
boreal rather than arctic species.*   It would thus appear
that, in the middle Pleistocene, land and fresh-water conditions prevailed in the region of the great lakes.

Dr. Frank D. Adams has recently made microscopical
examinations of specimens of the typical Erie clay from
the St. Clair tunnel, where it appears to be composed of
débris, both from the Laurentian cystalline rocks and the
Erian beds of the district.†

---

* Dawson and Penhallow, Pleistocene Flora of Canada, Bul. Am.
Geological Society, 1890 ;   Hinde, Interglacial Beds, *Canadian
Journal*, 1877.

† Trans. Royal Society of Canada, 1891.

The comparative table given in chapter II. shows that somewhat similar conditions prevail over the great plains

1. Section of Drift on Milk River—(*a*) Boulder-clay; (*b*) older river alluvium; (*c*) newer river alluvium; (*d*) surface matter.
2. Section of Drift at Long River—(*a*) Lower drift, chiefly local, and with false bedding (*b*) upper drift with large travelled boulders. [After Dr. G. M. Dawson.]

west of lake Superior, the formation of which have been described by Dr. G. M. Dawson in his Report on the 49th

Parallel and in his paper on the Superficial Deposits of the Plains. To these reference may be made for details.

The sections given in the figures represent some features of these deposits, and are interesting as showing its massive character in some places, and the fact of an underlying deposit of water-sorted material. The general structure, however, appears to be that stated in Chap. II., namely, an under and upper boulder-deposit, separated by beds of stratified silt, and sometimes holding vegetable matter.

I do not propose to extend these local details into the vast regions lying in the Arctic basin, north of the Laurentian water-shed and west of the basin of the great lakes in Manitoba and the North-west and in British Columbia. These regions have been described, the latter, from personal knowledge, and both, with reference to all the available authorities, by my son, Dr. G. M. Dawson, F.R.S., and I may refer to his two memoirs: " Notes on the Geology of the Northern Part of the Dominion of Canada," Reports Geological Survey of Canada, 1887; and " On the Later Physiographical Geology of the Rocky Mountain Region in Canada," Transactions of Royal Society of Canada, 1890. In these papers will also be found copious references to all previous explorations and sources of information.

EXPLANATION OF PLATE I.

The following plate, drawn under my own direction, is intended to present, as faithfully as possible, the characters of some of the more rare and critical shells of the Canadian Pleistocene.

Fig. 1. *Astarte Banksii*—A full-grown specimen of the ordinary type. Rivière-du-Loup.

Fig. 2. *Astarte Laurentiana* — An average full-grown specimen. Montreal.

Fig. 3. *Astarte lactea*—Ordinary type. Portland, Maine.

Fig. 4. *Astarte Elliptica*—A specimen with the ribs extending nearer to the ventral margin than usual. Portland, Maine.

Fig. 5. *Buccinum tenue*—Full-grown specimen. Rivière-du-Loup. 5a—Sculpture enlarged.

Fig. 6. *Buccinum cyaneum*—Full-grown specimen. Rivière-du-Loup. 6a—Sculpture enlarged.

Fig. 7. *Buccinum undulatum*—(Var. of *undatum*)—Immature shell, broken at lip. Rivière-du-Loup. 7a—Sculpture enlarged.

Fig. 8. *Buccinum glaciale*—Tuberculated variety. Rivière-du-Loup. 8a—Sculpture enlarged.

Fig. 9. *Buccinum glaciale*—Smooth variety. Rivière-du-Loup. 9a—Sculpture enlarged.

Fig. 10. *Buccinum ciliatum*—(Fabricius, not Gould)—Smooth variety, somewhat decorticated. Montreal. 10a—Sculpture enlarged.

Fig. 11. *Buccinum ciliatum*—(Fabricius, not Gould)—Small but mature specimen. Recent Murray Bay.

Fig. 12. *Buccinum Grœnlandicum*—Adult specimen. St. Nicholas. 12a—Sculpture enlarged.

Fig. 13. *Choristes elegans*—(Carpenter)—Adult specimen. Montreal. 13a—Sculpture enlarged.

Fig. 14. *Capulus commodus*—Pt. Levis, Quebec.

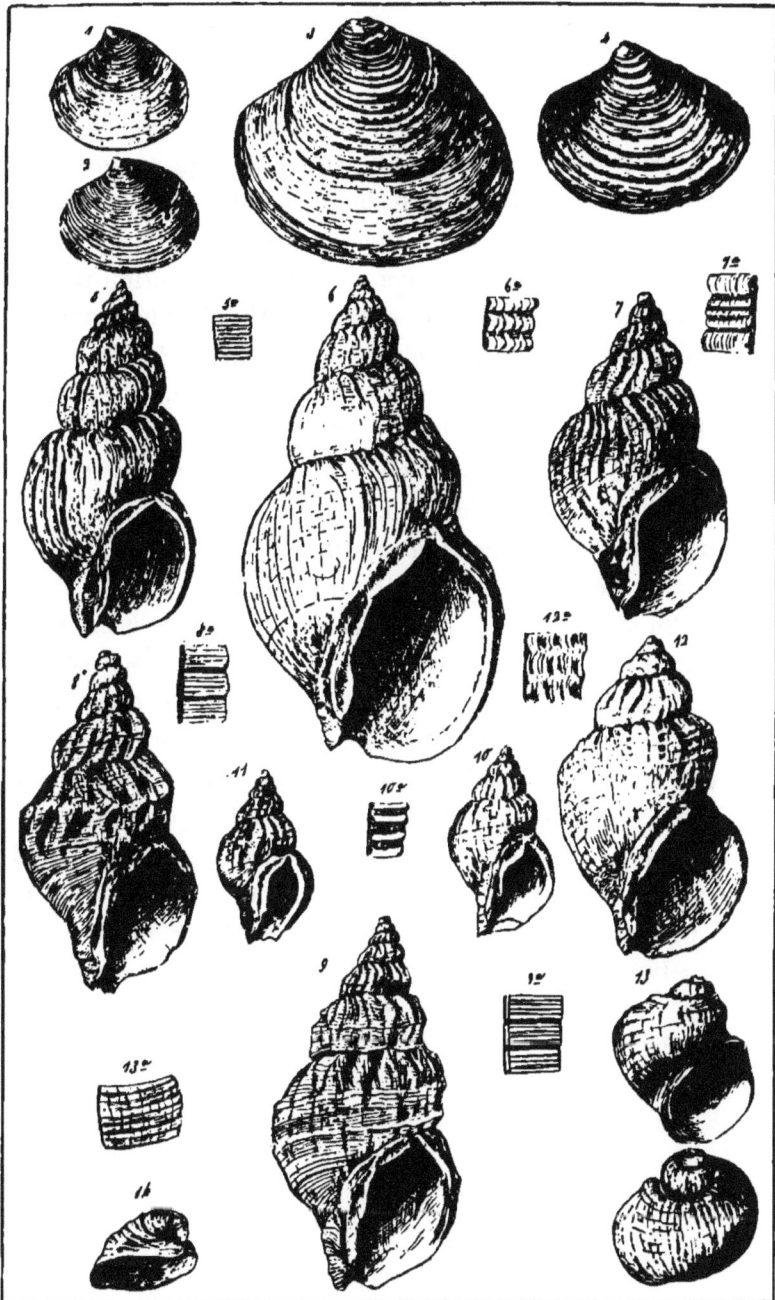

Characteristic Pleistocene Shells (see table of reference on previous page).

CHAPTER VI.

PLEISTOCENE FOSSILS.

This chapter is necessarily for reference rather than for reading. It represents, however, a large amount of patient work, and furnishes some of the most important data for conclusions as to the climate and physical conditions of the Pleistocene age. In this connection it will be observed that the greater part of the fossils recorded are from the St. Lawrence valley and the Atlantic coast, and from other areas in the arctic basin and west coast which were submerged in the Pleistocene; and that the evidences of life belong chiefly, though by no means exclusively, to the middle Pleistocene.

The lists of Pleistocene fossils of Canada, published previously to 1856 by Lyell and others, included only about 26 species. In my papers published between that year and 1863, the number was raised to nearly 80. These lists were tabulated, along with some additional species furnished in M.S., in the Report of the Geological Survey for 1863, the list there given amounting to 83 species, exclusive of Foraminifera. In my paper on the Post-pliocene of Rivière-du-Loup and Tadoussac, published in 1865, I added 38 species, and in the "Notes on the Post-pliocene of Canada" many others were introduced.

15

The number will be still further augmented in the following revision, which will afford a very complete view of the subject up to the present time; and though additional species will no doubt be found, yet all the principal deposits have been so carefully explored, that only very rare forms can have escaped observation. For some of the additional species included in the present list I am indebted to Prof. Kennedy (now of Windsor, N.S.), the late Dr. Anderson (of Quebec), the late Sheriff Dickson (of Kingston), Mr. T. Curry (of Montreal), Lieut.-Colonel Grant (of Hamilton), Dr. Packard, Mr. G. F. Matthew, Rev. Mr. Paisley, and other friends, to whom reference will be made in connection with the several species in the catalogue.

In so far as nomenclature is concerned, I have, wherever possible, retained the generic and specific names of the list published in 1872. Where errors had been committed, the names are of course changed, and any new generic names or possible identifications with other species are noted in brackets or otherwise.

In the case of the recent species of marine animals, those quoted are largely from my own dredgings in the Lower St. Lawrence, which I have prosecuted for many years with the view of ascertaining the modern habitats of the Pleistocene species; and reference is made to other collectors where advantage has been taken of their labours.

I am indebted to Mr. J. F. Whiteaves, F.G.S., paleontologist to the Geological Survey of Canada, for his kindness in looking over the list and adding some valuable corrections and suggestions. Mr. Lambe, of the Geological Survey, has also kindly examined the sponges, and given me his views as to their relations.

The fossils contained in the following lists have been presented to the Peter Redpath Museum of McGill University, Montreal, and are now exposed in its cases.

ANIMAL FOSSILS.

PROVINCE PROTOZOA.

(1) *Foraminifera.*

Nodosaria (Glandulina) lævigata.

———————————————(Var. *Dentalina communis*).

Fossil—Leda clay, Montreal.

Recent—Gulf St. Lawrence, 30 to 300 fathoms. G. M. D.*

This species is very rare in the Post-pliocene, but sometimes of large size and of different varietal forms.

Lagena sulcata —— (Var. *distoma*).

——————— (Var. *semisulcata*).

Fossil—Leda clay, Montreal ; Quebec ; Murray Bay ; Anticosti ; Rivière-du-Loup ; Portland (Maine).

Recent—Gulf St. Lawrence, 18 to 313 fathoms. G. M. D. British Columbia.†

Rather rare in the Pleistocene as well as in the recent.

Entosolenia globosa.

————*costata.*

————*marginata.*

————*squamosa.*

Fossil—Montreal, Leda clay ; Labrador ; Rivière-du-Loup ; Anticosti ; Murray Bay ; Quebec ; Portland (Maine).

Recent—Gulf and River St. Lawrence, 20 to 313 fathoms. G.M.D.

Generally diffused in the Pleistocene, and presenting the same range of forms as in the recent ; but not common. I regard the supposed species of *Entosolenia* above named as merely varietal forms.

———————————————————————————————

* The initials G. M. D., refer to the List of Foraminifera by Dr. G. M. Dawson in *The Canadian Naturalist*, 1870.

† This and other species from British Columbia are from a memoir by Whiteaves on collections of Dr. G. M. Dawson. Trans. R. S. Can., Vol. IV., 1887.

Bulimina Presli.

——————— (Var. *squamosa*).

Fossil—Montreal, Leda clay; Labrador; Rivière-du-Loup; Murray Bay; Quebec; Portland (Maine).

Recent—Gulf and River St. Lawrence, 10 to 313 fathoms.   G. M. D.

FOSSILS—PLATE II.

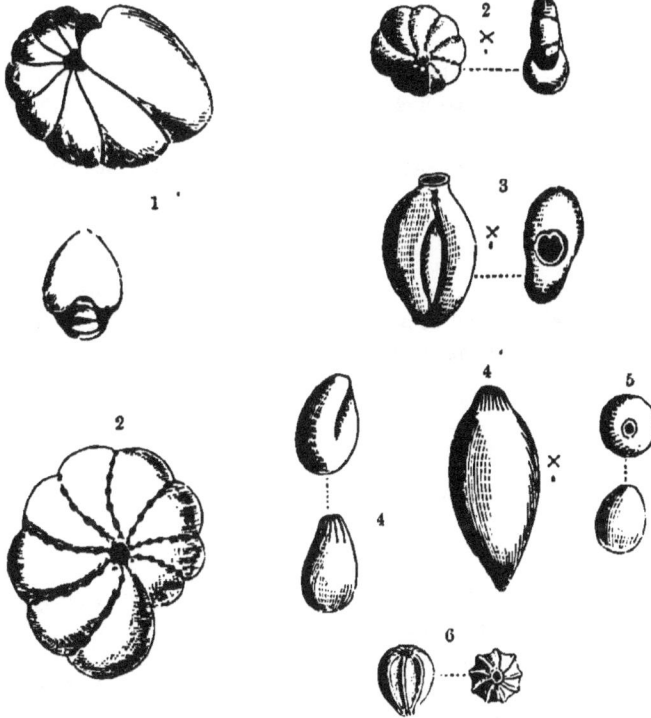

Pleistocene Foraminifera.—1, *Nonionina scapha* (Var. *Labradorica*) ; 2, *Polystomella umbilicatula;* 3, *Quinqueloculina seminulum;* 4, *Polymorphina lactea* (2 varieties) ; 5, *Entosolenia globosa ;* 6, *E. costata.*   (All magnified.)

Generally diffused in the Pleistocene.   In the recent it seems to be mostly a deep-water form.   What Parker and Jones call the essentially arctic form *B. elegantissima* is not uncommon, though other forms also occur.

Polymorphina lactea.

Fossil—Montreal, Leda clay ; Labrador ; Rivière-du-Loup; Murray Bay.

Recent—Gulf and River St. Lawrence, 30 to 313 fathoms. G. M. D.
British Columbia.

Not uncommon in the Pleistocene, particularly in the deeper part
of the Leda clay. Less common recent. I observed in the Rivière-
du-Loup gatherings a small individual of this species with the internal
pipe at the aperture characteristic of Entosolenia, which is also sometimes observed in recent specimens.

Truncatulina lobatula.

Fossil—Leda clay, Labrador ; Rivière-du-Loup ; Anticosti.
Recent—Gulf St. Lawrence, very common 30 to 50 fathoms. British
Columbia.

This species is much less common in the Pleistocene than in the
recent.

Orbulina universa.

Fossil—Leda clay, Montreal ; Rivière-du-Loup ; Labrador.
This may be regarded as a rare and somewhat doubtful Pleistocene
fossil. It has not yet been recognized in the Gulf of St. Lawrence.

Globigerina bulloides.

Fossil—Rivière-du-Loup ; Anticosti.
Recent—Gulf St. Lawrence, more especially in the deeper water,
where it is common. It is very rare in the Pleistocene.

Pulvinulina repanda.

Fossil—Montreal, Leda clay ; Rivière-du-Loup ; Murray Bay ;
Labrador ; Quebec ; Portland (Maine).
Recent—Gulf St. Lawrence, 30 to 313 fathoms. G. M. D.
Somewhat rare both in the Pleistocene and recent, and of the small
size usual in the arctic seas.

Polystomella crispa.—(Var. *striatopunctata*).
——— ——— ——— (Var. *arctica*).

Fossil—Montreal, Leda clay ; Labrador ; Rivière-du-Loup ; Murray
Bay ; Quebec ; Portland (Maine) ; St. John, N.B.
Recent—Gulf and River St. Lawrence, 30 to 40 fathoms. G. M. D.
British Columbia.

Very common, especially in depths of 10 to 40 fathoms. This is by
far the most abundant species in the Pleistocene deposits, as it is also
in all the shallow parts of the Gulf of St. Lawrence at present, and

also in the Arctic Seas, according to Parker and Jones. It is the only species yet found in the Boulder-clay of Montreal, and this very rarely.

Nonionina scapha.
———————— (Var. *Labradorica*).

Fossil—Leda clay, Montreal ; Rivière-du-Loup ; Anticosti ; Labrador ; Murray Bay ; Quebec ; St. John, N.B.

Recent—Gulf and River St. Lawrence, 10 to 313 fathoms. (G. M. D.) . British Columbia. Var. *Labradorica* is the deeper water form and is rare in the Leda clay.

Textularia pygmœa.

Fossil—Leda clay, Labrador ; Rivière-du-Loup ; Quebec ; also at Portland (Maine).

Recent—Gulf St. Lawrence, 10 to 30 fathoms.

The Textulariæ are rare and of small size, both in the Pleistocene and recent.

Cornuspira foliacea.

Fossil—Leda clay, Montreal.

Recent—Gulf St. Lawrence, 16 to 250 fathoms. G. M. D.

This species is rare both fossil and recent.

Quinqueloculina (Miliolina) seminulum.

Fossil—Leda clay, Montreal; Labrador ; Quebec; Portland (Maine). British Columbia.

Recent—Gulf St. Lawrence, 10 to 313 fathoms, most abundant in shallow water. G. M. D.

This species is by no means common and not usually large in the Pleistocene. It is more abundant in the clays of Maine than in those of Canada.

Biloculina ringens.

Fossil—Leda clay, Montreal; Labrador; Rivière-du-Loup ; Murray Bay ; Quebec.

Recent—Gulf St. Lawrence, 30 to 313 fathoms. G. M. D.

Rather rare in the Pleistocene as well as in the recent.

Triloculina tricarinata.

Fossil—Leda clay, Rivière-du-Loup ; Murray Bay ; Quebec.

Recent—Gaspé, 30 to 50 fathoms. G. M. D.

Rare both in Pleistcene and recent, but perhaps more generally diffused in the former.

Lituola and *Saccammina.*

A very few minute sandy forms referable to these genera are found among the finer parts of the washings from Rivière-du-Loup.

Euglypha?

A single minute test, apparently identical in form with that of *Euglypha alveolata*, was found in washing the Rivière-du-Loup clays.

In general terms it may be stated that all the species of Foraminifera found in the Pleistocene still inhabit the gulf and river St. Lawrence. Several species found in the gulf of St. Lawrence have not yet been recognized in the Pleistocene, and these are mostly inhabitants of depths exceeding 90 fathoms, or among the more southern forms found in the gulf.

On the whole, the assemblage, as in the northern part of the gulf of St. Lawrence at present, is essentially arctic, and not indicative of depths greater than 100 fathoms, which would seem to have been the maximum depth of the sea of the Leda clay, and corresponds with well-marked terraces on the hills.

The sandy forms, which are not uncommon in the Gulf, are very rare in the Pleistocene; but this may be accounted for by the greater difficulty of washing them out of the clay, or possibly their cementing material may have decomposed, allowing them to fall to pieces. As the epidermal matter of shells is often preserved, the last supposition seems less likely. The Leda clays are, however, usually very fine and calcareous, so that there was probably more material for calcareous than for arenaceous forms.

The Foraminifera are very generally diffused in the Pleistocene clays, though much more abundant in some layers than in others. They may easily be detected by a pocket lens, and are usually in as fine preservation as recent specimens, especially in the deeper and more tenacious layers of the Leda clay. They are, however, usually more abundant in the somewhat arenaceous layers near the top of the Leda clay, and immediately below the Saxicava sand, and especially where this layer contains abundance of shells of mollusca. I have nowhere found them more abundant or in greater variety than at the Glen brick-work near Montreal, on the McGill College grounds, and at Logan's Farm., At the Glen brick-work a few worn specimens of Polystomella are contained in the beds underlying the Leda clay and equivalent to the boulder-clay, which, however, has in general, in the vicinity of Montreal, as yet afforded no marine fossils.

In searching for Foraminifera in the clays of Rivière-du-Loup, I have observed in the finer washings several species of Diatomaceæ; and among these a species of *Coscinodiscus* very frequent in the deeper parts of the gulf of St. Lawrence. But on the whole diatoms appear to be rare in these deposits. In the Rivière-du-Loup clays I have also observed the pollen grains of firs and spruces.

The nomenclature used above is that of Parker and Jones, in their paper on the North Atlantic Soundings, in the Transactions of the Royal Society. For figures of the species, I may refer to that memoir, and to my previous papers published in the *Canadian Naturalist*.

(2) *Porifera*.

Tethea Logani. Dawson.

Leda clay, Montreal. This species has not yet been recognized in a living state. Its spicules in considerable masses, looking like white fibres, are not uncommon in the Pleistocene at Montreal.

According to Mr. Lambe, this species is a tetractinellid sponge; but possibly different forms have inadvertently been included under one name.

FOSSILS—PLATE III.

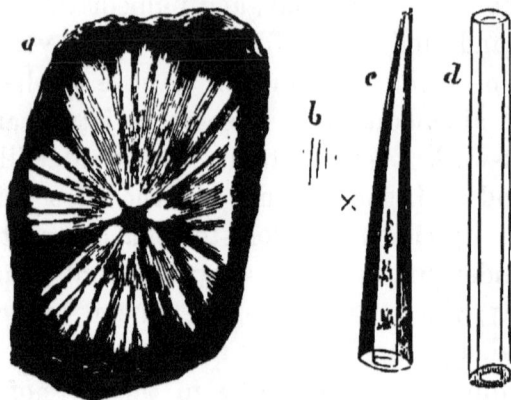

Tethea Logani.—(*a*) Mass of radiating spicules in clay ; (*b*) Spicules ; (*c, d*) Portions of spicules enlarged. *Ophioglypha Sarsii.*

Tethea ?

Another silicious sponge is indicated by little groups of small spicules found at the Tanneries, near Montreal, by Mr. G. T. Kennedy, and at Rivière-du-Loup by the author. Its spicules are long and acerate, and much more slender than those of *Tethea Logani.*

PROVINCE CŒLENTERATA.

Hydrozoa.

No distinct organisms referable to the above group have yet been found in the Pleistocene deposits of Canada. As our recent fauna includes no stony coral, and the recent species of the gulf of St. Lawrence have no parts likely to be preserved other than minute spicules, this is not to be wondered at. In washing the clays for Foraminifera, however, numerous fragments are obtained, which resemble portions of the horny skeletons of hydroids, though not in a state admitting of determination.

PROVINCE ECHINODERMATA.

(1) *Ophiuridea.*

Ophioglypha Sarsii. **Lutken.**

Fossil—Leda clay, near St. John, New Brunswick ; Mr. Matthew.

Recent—River St. Lawrence, at Murray Bay ; Kamouraska ; also found of large size in deep water in the Gulf of St. Lawrence, by Mr. Whiteaves.

Ophiacantha Spinulosa, M. and T. (*O. bidentata,* Retzius).

Fossil—Tanneries, collected by Prof. Kennedy and Mr. Currie.

Recent—Cape Cod to Greenland, Norway and Spitzbergen.

Amphiura Sp. Montreal, Mr. Currie.

Solaster (Crossaster) papposa, M. and T.

Fossil—Montreal, Mr. Kennedy ; Green's Creek.

Recent—Labrador, Murray Bay, Metis, Gaspé.

Ophiocoma or *Amphiura.*

Fragments of a small species of ophiuroid starfish not determinable, have been found in the Leda clay at Montreal, and in nodules at Green's creek.

(2) *Echinoidea.*

Euryechinus (Strongylocentrotus) drobachiensis. **Müller.**

Fossil—Leda clay, Beauport ; Rivière-du-Loup ; St. Nicholas ; Montreal.

This species is rare in the Pleistocene, but very common in all parts of the gulf of St. Lawrence at present. Also west coast British Columbia.*

(3) *Holothuridea*.

Psolus (*Lophothuria*) *Fabricii*, Dur. and Kor.

Scales of an animal of this kind have been found in the Leda clay at Montreal. They may belong to *P. phantopus*, or to the species *P.* (*Lophothuria*) *Fabricii*, also found on our coasts ; from the form of the plates, I suppose most likely to the latter species, and to a young or small individual. *P. Fabricii* is very abundant at Little Metis, where the other species also occurs.

PROVINCE MOLLUSCA.

Introductory.—In preparing this, the largest and most important part of my catalogue, I have to acknowledge my obligations to the late Dr. P. P. Carpenter, for his kind aid in comparing all the more critical species of shells, and in giving me his valuable judgment as to their relations and synonymy, which I have in nearly every case accepted as final. I am also indebted to Dr. Carpenter for many of the notices of West-coast shells.

To Mr. J. F. Whiteaves, F.G.S., I am indebted for reviewing the Polyzoa and comparing them with Smitt's Norwegian catalogues, and also for many valuable facts as to shells obtained in his dredgings in the gulf of St. Lawrence. The Rev. T. Hincks, F.R.S., has also given me valuable information on the Polyzoa.

To the late Mr. J. Gwyn Jeffreys, F.R.S., and the late Mr. R. McAndrew, F.R.S., of London, my grateful acknowledgments are due for aid and information, and also for

---

* Names of recent shells from British Columbia, except when otherwise credited, are quoted from papers of Mr. Whiteaves on collections of Dr. G. M. Dawson ; Trans. R. S. Canada, Vol. IV., 1886; Geol. Survey of Canada, 1878-9.

the opportunity of comparing my specimens with those in their collections.

My comparisons with recent species have been made to a great extent with specimens dredged by myself, in the gulf and river St. Lawrence, and especially at Murray bay and Metis, where the marine fauna seems to be more nearly related to that of the Pleistocene than in any part of the gulf of St. Lawrence with which I am acquainted. I have also to acknowledge the use of specimens from Greenland, from Prof. Morch; from Norway from Mr. McAndrew; from Nova Scotia from Mr. Willis; as well as the use of the large and valuable collections of Dr. Carpenter and Mr. Whiteaves.

All the references in the following pages, except where authors are quoted, and many of these last, have been verified by myself by actual comparison of specimens.

The principal works to which I have referred in the publication of the catalogue are the following:

Beechey's Voyage, Natural History Appendix.
Belcher's Last of the Arctic Voyages, do.
Bell, Report on Invertebrata of Gulf of St. Lawrence.
Busk, Polyzoa of the Crag.
Crosskey on Post-pliocene of Scotland.
Fabricius, Fauna Grœnlandica.
Forbes and Hanley, British Mollusca.
Gould, Invertebrata of Massachusetts, edited by Binney.
Jeffreys' British Conchology.
Lyell on Shells collected by Captain Bayfield; and Travels in North America.
Matthew on Post-pliocene of New Brunswick.
Middendorff, Shells of Siberia.
Packard on the Glacial Phenomena of Labrador and Maine.
Prestwich on the English Crag.
Sars on the Quaternary of Norway.
Stimpson, Shells of Hayes' Expedition, &c.

Whiteaves, Lists of Shells from Gulf of St. Lawrence, *Canadian Naturalist*.

Wood, Crag Mollusca.

FOSSILS—PLATE IV.

1

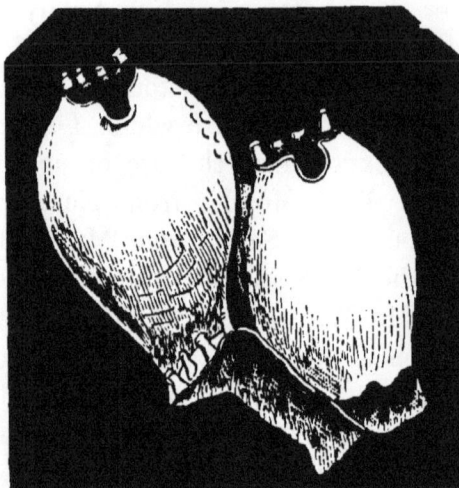

2

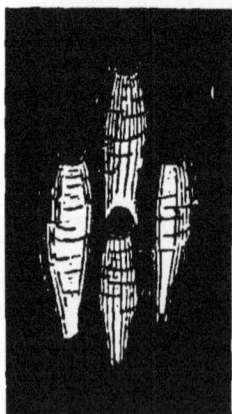 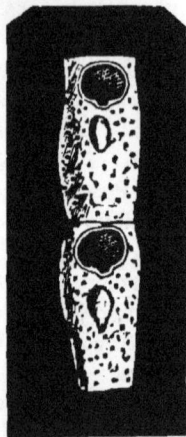

Pleistocene Polyzoa (magnified).—1, *Lepralia quadricornuta*; 2, *L. hyalina* and *L. pertusa.*

Willis, Lists of Shells of Nova Scotia.

Also various recent papers by Jeffreys, Hincks, Brady, Verrill, Sars and others.

CLASS I.—MOLLUSCOIDEA.

Sub-Class 1.—*Polyzoa.*

Membranipora catenularia. Jameson.

Fossil—Beauport ; Labrador ; Rivière-du-Loup.
Recent—Gaspé ; * Labrador (Packard).

Membranipora Lacroixii. Audouin.

Fossil—Rivière-du-Loup.
Recent—Gaspé ; Labrador (Packard).
Entirely agrees with recent examples from Gulf of St. Lawrence.

Membranipora lineata. Linn.

Fossil—Rivière-du-Loup.
Recent—Gaspé.

Membranipora pilosa. Linn.

St. John (Matthew).

Hippothoa expansa, Dawson.

Fossil—Beauport ; Rivière-du-Loup.
Recent—Gaspé ; Labrador ; Maine (Packard).

Schizoporella hyalina. Johnston.

Fossil—Beauport ; Rivière-du-Loup ; St. John (Matthew).
Recent—Gaspé.

Lepralia pertusa. Esper.

Fossil—Beauport ; Labrador ; Rivière-du-Loup.
Recent—Gaspé ; Labrador (Packard).

Lepralia quadricornuta. Dawson.

Fossil—Leda clay, Montreal.
Not yet found recent.

Mr. T. Curry, of Montreal, has recently found specimens in a very perfect state. They show that the cells are sculptured in a papillo-striate manner, and that the ovi-capsules are globular and granulate. Some cells have a projection for a vibraculum or avicularium at one

---

* The references to Gaspé are from my list contributed to the Rept. Geol. Survey, 1858—Bell and Richardson, collectors ; and from subsequent dredgings by myself and Mr. Whiteaves.

side of the aperture A few have two of these. Old colonies have a pitted calcareous deposit between the cells. The large size and narrow aperture with deep sulcus in front and four spines behind are as in the specimens formerly described.

Specimens of this species were sent in 1883 to Rev. Thomas Hincks, and he writes to me as follows : " The specimens of *Lepralia quadri-cornuta* which you were kind enough to send me reached me in perfect safety, and I have been much interested in examining them. None of the specimens that I had seen previously showed the real character of the orifice, and I was therefore led to refer your species to *L. coccinea*, which, in most points of structure, it closely resembles. Indeed, the only difference of much significance is found in the form of the mouth, but this is very marked and distinctive. I am now fully convinced that your species is a good one, and shall take an early opportunity of correcting the error in my 'History.' It has a special interest as being the only Post-pliocene form that is not known to have survived."

Lepralia spinifera? Busk.

Fossil—Rivière-du-Loup.

L. Violacea? Johnston.

Fossil—Rivière-du-Loup.

L. variolosa. Johnston.

Fossil—Rivière-du-Loup.
Recent—Gaspé.

L. globifera. Packard.

Fossil—Rivière-du-Loup.
Recent—Labrador (Packard).

Porella Belli. (*Lepralia Belli.* Dawson.)

Fossil—Rivière-du-Loup.
Recent—Gaspé ; Labrador (Packard).
This species, according to Hincks, is identical with *Porella Coricinna.* Busk.

Porella elegantula. D'Orbigny.

Fossil—Rivière-du-Loup ; Montreal (Curry).
Recent—Labrador (Packard) ; Gaspé.
Very fine and frequent in 10-30 fathoms opposite Cape Rosier Village. J. F. W.

Porella compressa. Sowerby. (= *Celleporaria surcularis.* Packard.)
Fossil—Rivière-du-Loup.
Recent—Labrador (Packard) ; Gaspé.
Abundant in 10-50 fathoms everywhere in the Gulf, and often drifted down to lower levels. J. F. W.

Smittia producta. Packard. (Sp.)
Fossil—Rivière-du-Loup.
Recent—Labrador (Packard) ; Gaspé ; Murray Bay.

Smittia trispinosa. Johnston.
Fossil—Rivière-du-Loup.
Recent—Gaspé ; Labrador (Packard).

Cribrilina punctata? Hassall.
Fossil—Rivière-du-Loup.
Gaspé.

Mucronella Peachii. Johnston.
Fossil—Rivière-du-Loup.
Recent—Gaspé.

Mucronella ventricosa. Hassall.
Fossil—Rivière-du-Loup.
Recent—Gulf St. Lawrence.

Myriozoum planum. Dawson. (Sp.)
Fossil—Rivière-du-Loup.
Recent—Gaspé.

Myriozoum sub-gracile. D'Orbigny.
Fossil—Rivière-du-Loup.
Recent—Labrador (Packard) ; Gaspé.

Cellepora pumicosa. Ellis.
St. John (Matthew).

Tubulipora fimbria. Lam. (= *T. flabellaris.* Johnston, non Fabr.)
Fossil—Beauport ; Rivière-du-Loup.
Recent—Gaspé, Labrador (Packard) (= *T. Palmata,* Wood).

Idmonea atlantica. Forbes.
Fossil—Rivière-du-Loup.
Recent—I believe this to be identical with the species found in the gulf of St. Lawrence, and referred by Dr. Packard and Mr. Whiteaves to the above.

Diastopora obelia. Johnston.

Fossil—Rivière-du-Loup.

Recent—Gaspé.

Crisia eburnea. Ellis.

Fossil—Montreal. A specimen collected by Mr. Curry is referred to this species by Mr. Whiteaves.

Recent—Labrador (Packard). In 96 fathoms, Trinity Bay, N. Shore St. Lawrence R. J. F. W.

Alecto. (Sp.)

Fossil—Rivière-du-Loup.

Lichenopora hispida. Fleming. (Sp.)

Fossil—Rivière-du-Loup. Patches on shells, somewhat worn, but referable to this common North Atlantic species.

2.—Sub-Class—*Brachiopoda.*

Rhynchonella psittacea. Gm.

Fossil—Montreal; Beauport; Anticosti; Rivière-du-Loup. Abundant.

Recent—Murray Bay and Gaspé; Little Metis; Kamouraska; British Columbia; Labrador (Packard); Gulf St. Lawrence. Generally on stony bottoms 10 fathoms and over. Arctic seas generally; also Crag of England and glacial beds.

In a bed of stony clay at Rivière-du-Loup, this shell is very abundant, with less abundant specimens of the next species. It occurs living in precisely the same relations and in great abundance at Murray Bay, in about 20 fathoms.

Pleistocene Brachiopods.—1. Rhynchonella psittacea. 2. Terebratella Spitzbergensis.

Terebratella Spitzbergensis. Davidson.

Fossil—Rivière-du-Loup.

Recent—Murray Bay; also at several localities in the gulf of St. Lawrence (Whiteaves); Nova Scotia (Willis).

This species has been found in the Pleistocene of Canada, hitherto only at Rivière-du-Loup, and is rare. It appears to be a rare shell in every part of the Gulf where it has hitherto occurred, except at Murray bay, where it is not uncommon, and is found attached to stones in 20 to 25 fathoms, associated with *Rhynchonella psittacea.*

16

FOSSILS—PLATE V.

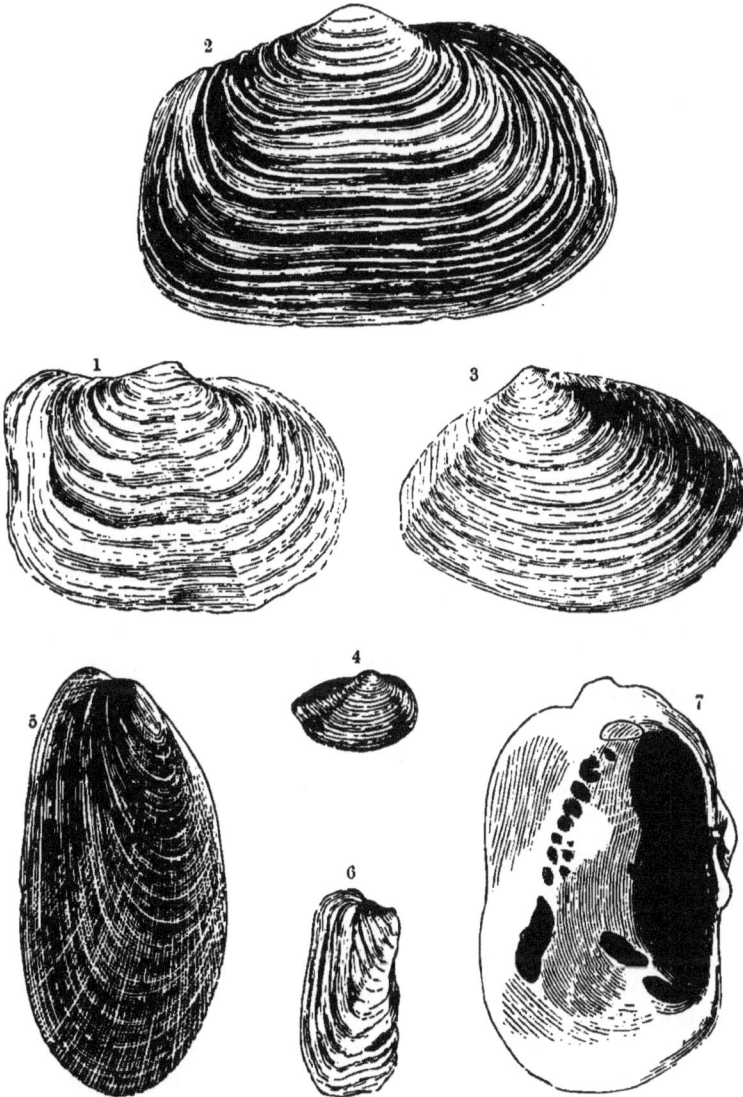

Pleistocene Lamellibranchs.—1, *Mya trunca'a*—Var. *Uddevallensis*—Montreal ; 2, *Mya truncata*—Var. *communis*—Portland ; 3, *Macoma calcarea ;* 4, *Leda arctica ;* 5, *Modiolaria nigra ;* 6, *Saxicava rugosa*—Var. *arctica*—Montreal ; 7, *Panopea Norvegica,* Rivière-du-Loup.

CLASS II.—LAMELLIBRANCHIATA.

Pholas (Zirphea) crispata. Linn.

Fossil—Maine (Packard).

I have not found this species fossil in Canada, but it exists as a living shell on the New England coast generally, in Northumberland strait; gulf of St. Lawrence, and, according to Bell, as far to the north-west as Rimouski. Puget Sound (U. S. Expl. Exped.) Queen Charlotte Islands (Whiteaves).

It has perhaps extended its northern limit to Canada since the glacial period. On the European coast it is a northern shell, reaching south to the Mediterranean.

Saxicava rugosa. Lamarck (and var. *arctica*).

Fossil—Saxicava sand and top of Leda clay, Montreal; St. Nicholas, Ottawa; L'Orignal; Chaudière Station; Upton, P.Q.; Stormont, Ont.; Quebec; Murray Bay; Rivière-du-Loup; Trois Pistoles; Tadoussac; Anticosti; Labrador; Lawlor's Lake; Bathurst; New Richmond; * Vancouver Island (G. M. Dawson); New Brunswick; Maine, &c.

Recent—Gulf St. Lawrence; coast of Nova Scotia; and New England and northern seas generally; also British Columbia and west coast of America as far as Mazatlan. (P. P. Carpenter).

Very abundant in the more shallow-water portions of the Pleistocene throughout Canada, and presenting all the numerous varietal forms of the species in great perfection. It is relatively much more abundant in the drift-deposits than in the gulf of St. Lawrence at present. Pieces of limestone which have been bored probably by this shell, are not rare in the drift at Montreal.

This is a widely distributed arctic species, and is found in the Pleistocene deposits of Europe, and as far back as the Miocene.

Panopœa Norvegica. Spengler.

Fossil—Leda clay; Rivière-du-Loup. Very rare.

Recent—Little Metis; dredged in Gaspé Bay, 30 and 40 fathoms, by Mr. Whiteaves; Halifax (Willis); Grand Manan (Stimpson); Arctic and northern seas generally.

It is very rare in the Pleistocene, a few valves only having been found at Rivière-du-Loup. The specimens are small, and much inferior to those found in the Scottish Clyde beds, of which I have a specimen from Rev. H. Crosskey.

* For shells from this locality I am indebted to Dr. Thornton, of New Richmond.

Mya truncata. Linn. (And var. *Uddevallensis.*)

Fossil—Saxicava sand and Leda clay; Montreal; Quebec; Rivière-du-Loup; Anticosti; Goose River, N. Shore, River St. Lawrence; New Richmond; Portland; New Brunswick (Matthew); Labrador (Packard); Greenland (Möller); also in the Pleistocene of Europe.

Recent—Little Metis; Tadoussac; Rivière-du-Loup; British Columbia; Gulf St. Lawrence, but rare in comparison with its abundance in the drift. Generally distributed in the Arctic seas and North Atlantic, American coast as far south as Cape Cod; Puget Sound (= *preciosa* Gould, P. P. C.)

The variety usually found in the Pleistocene of Canada is the short or *Uddevallensis* variety, which is that occurring in the arctic seas at present, while in the Gulf St. Lawrence the ordinary long variety is found almost exclusively. At Portland, Maine, however, the long variety lived in the Pleistocene, and occasional specimens are found at Rivière-du-Loup and New Richmond. The form *Uddevallensis* occurs living in Labrador (Packard), and I have found it at Tadoussac and Little Metis.

It is interesting to observe that while the present species is more abundant than the next in the Pleistocene, it is much more rare in the Gulf at present. It also occurs in deeper water.

In collecting recent specimens of *Mya truncata* and *M. arenaria* at Little Metis, I have had opportunity to observe their habits and varieties in a manner to illustrate the differences above noticed.

At the head of Little Metis Bay, where the water is shallow and warm, and the bottom is soft mud and sand, a large variety of *Mya arenaria* is very plentiful in the flats bare at low tide; so much so that the place is resorted to by fishermen from localities lower on the coast for bait. It sometimes attains the length of $4\frac{1}{2}$ inches, and has a thick, dense shell, without perceptible epidermis, and often with radiating bands. So far as I am aware, neither *Mya truncata* nor the peculiar variety of *M. arenaria* referred to below, occurs on this part of the coast.

I have not infrequently dredged *Mya truncata*, usually the long variety, but sometimes the short Uddevalensis variety, in deep water outside the bay, but have not seen it above low-water mark, though it occurs not far from this line; and, on the opposite side of the river St. Lawrence, I have found it at Tadoussac, where the water is still colder, close to low-water mark. I was not aware till lately that *Mya arenaria* occurred on the comparatively steep and stony shore outside the bay, and it is certainly not found there inside of the low-water limit.

In 1888, however, after a heavy easterly gale, great numbers of *Mya arenaria*, in a living state, and a few specimens of *M. truncata*, were thrown up on the beach, and must have been derived from the mud disturbed by the breakers at no great distance outside of low-water mark, or on a slight bank a little further seaward. The former were all of small or moderate size, somewhat round and flat in form, much wrinkled and covered with a thick brown epidermis which extended a little way beyond the posterior end of the shell, which was, however, rounded and not truncated, and destitute of the corneous tube of *M. truncata*. Still, many of the specimens might, at first sight have been mistaken for *M. truncata*, with the tube partly broken off. This enabled me, for the first time, to understand the remark of Fabricius, that in Greenland the two species are so similar, that but for the hinge and the tube they might be confounded. With these were thrown up specimens of *M. truncata*, which must have lived with the others, the inner limit of *M. truncata* probably overlapping the outer limit of *M. arenaria*. The short or Uddevalensis variety of *truncata* was, however, very rare, only a few shells in a perfectly recent state having been found, and they probably lived in somewhat deeper and colder water than the others. The water, I may add, on this coast is so far affected by the arctic current as to be quite cold, except near the shore and in shallow bays, and the species dredged in 10 to 15 fathoms are, in general, similar to those of the Labrador coast, belonging rather to the boreal than to the Acadian fauna. With the Myas were cast up shells of *Solenensis*, var. *Americanus* of Carpenter, and of *Machaera costata*, the latter sometimes of large size, though it is more abundant in the warmer water at the head of the bay, where *Purpura Lapillus*, a rare shell on this coast, also occurs on the reefs.

It is evident that though there is no passage from one species into the other, the long variety of *Mya truncata* represents the extreme limit of modification of that species for a shallow and warm-water habitat, while the small epidermis-clad variety of *M. arenaria* represents its extreme modification for deeper and colder water than usual; and along the coast at Metis these two varieties meet.

The coldness of the Pleistocene seas thus explains the occurrence, in the upper Leda clay, of the peculiar small and epidermis-clad variety of *M. arenaria* and of the short form of *Mya truncata*. The conditions in the colder parts of the river St. Lawrence approach in these respects to those of the Pleistocene, though they are no doubt more fully realized in the arctic seas.

As I have remarked in my notes on the Post-pliocene, the brown wrinkled epidermis-clad variety of *M. arenaria* occurs plentifully along with *M. Uddevalensis* in the upper Leda clay at Rivière-du-Loup.

From the accounts of arctic collectors from Fabricius downwards, it would appear that in Greenland, as in

Pleistocene Canada, *M. truncata* is very abundant, and occurs at low water in the sands, as *M. arenaria* does further south. It would seem also that it forms a large part of the food of the walrus and other animals, and is much used by the inhabitants. It also appears that a small variety of *M. arenaria*, with brown epidermis, is most common in Greenland, and occurs with *Mya truncata*, which is, however, more plentiful. The description given by Fabricius of *M. arenaria* obviously agrees with that of my small and brown variety from Metis.

It is interesting to note the companionship of these allied species in the North Atlantic throughout the Pleistocene and Modern periods, and the range of varietal forms applicable to each, according to the conditions to which they have been exposed, along with their continued specific distinctness, and the preference of each for certain kinds of environment, so that in some places one, and in others the other, predominates, while this relative predominance, as well as the prevalence of certain varietal forms, might no doubt be reversed by change of climate or of depth.

Mya arenaria. Linn.

Fossil—Leda clay and lower part of Saxicava sand; Montreal; Upton; Quebec; Murray Bay; Labrador; Duck cove and Lawlor's lake, New Brunswick (Matthew); Anticosti; Goose River; New Richmond; Tatagouche River, N.B. (Paisley); Gardiner, Maine; Upton, P.Q.; Portland, Maine; Greenland (Möller); also in the Post-pliocene of Europe.

Recent—Little Metis; Rivière-du-Loup, &c. Very abundant throughout the Gulf of St. Lawrence and coast of Nova Scotia and New England, also Arctic seas generally. Mr. Jeffreys considers it identical with *M. Japonica.* Jay. This or allied in W. America, P. P. C.

In the Gulf this species grows to a large size; I have a specimen five inches long from Gaspé; but in the Post-pliocene it is small and often of a short and rounded variety. This is especially the case inland, as at Montreal. At Rivière-du-Loup a small thin variety with a strong epidermis and attenuated posteriorly, is found *in situ* in its burrows in the Leda clay. It is a deep-water variety. Some large specimens in collections from this place, I have reason to believe, are from aboriginal kitchen-middens.

Pandora (Kennerlia) glacialis. Leach.

Fossil—Leda clay; St. John, New Brunswick (Matthew) ; Saco, Maine.

Recent—Gaspé (Whiteaves) ; Murray Bay ; Labrador (Packard) ; Little Metis.

This species, which was at first confounded with *Pandora trilineata*, is apparently quite distinct, and on the evidence of the hinge would seem to belong to a different genus. Much nearer to *Pandora pinna*, Mont. ; = *P. obtusa* Forbes and Hanley. J. F. W. Jeffreys regards these as varieties of *P. inequivalvis*.

Lyonsia arenosa. Möller.

Fossil—Leda clay ; Montreal (rare and small) ; Rivière-du-Loup, common ; Duck Cove, N.B. (Matthew) ; Saco, Maine ; also in Greenland (Möller).

Recent—Murray Bay, Rivière-du-Loup, Little Metis and Gaspé : Halifax (Willis) ; Greenland (Möller) ; Labrador (Packard).

Some specimens from Portland are much larger than those from Rivière-du-Loup and Montreal, and Mr. Whiteaves finds individuals an inch long, living at Gaspé.

Thracia Conradi. Couthuoy.

Fossil—Saco (Packard).

Not yet found fossil in Canada, but recent, though rare, in Nova Scotia (Willis) ; and at Gaspé. Abundant and large in Shediac Bay (J. F. W.) Also, though apparently rare, in Labrador (Packard).

Has probably extended its northern limit to Canada, since the glacial period.

Macoma Grœnlandica. Beck. (*Macoma fragilis* Fabr. sp. J. F.W.)

Fossil—Saxicava sand and Leda clay ; Montreal; Ottawa ; Perth, Ont. ; Pakenham Mills, Cornwall; Clarenceville ; Upton ; Quebec ; Murray Bay ; Rivière-du-Loup ; Chaudière Station ; Anticosti ; Upton, P.Q. ; Stormont, Ont. ; Labrador ; Lawlor's lake and Bathurst, N.B. ; Campbellton, P.E.I. ; Westbeach, Maine ; Greenland (Möller).

Recent—Everywhere on the coasts of the gulf and river St. Lawrence, as a common littoral shell. I have found it as far up the river as Kamouraska.

A thin and delicate variety with smooth epidermis is found in the Leda clay; coarser and more wrinkled varieties in the Saxicava sand. Larger specimens are found at Quebec and Rivière-du-Loup than more inland.

In the modern Gulf, the small and depauperated varieties are littoral and near the brackish water, the finer varieties passing into *Macoma fusca* of Say, which is a southern variety, are found on the coast of Nova Scotia and in the bay of Fundy. This shell is represented in the European seas and Post-pliocene deposits by the closely allied species *M. so'idula* or *Balthica*, which seems to pass through a corresponding series of varieties, but to be distinct. On the western American coast it is similarly represented by *M. inconspicua*. Mr. Tryon and Mr. Whiteaves believe the three forms to be conspecific.

It is said to be the *Tellina Fabricii* of Hanley, and I have specimens from Greenland from Morch labelled *T. tenera*. The *T. tenera* of Leach, however, is *proxima*, Brown, teste Hanley. It is apparently the *Venus fragilis* of Fabricius.

It is one of the most common and abundant shells of the Pleistocene, as it is of the American coast from Greenland to New England.

Macoma calcarea. Chemnitz.

Fossil—Leda and boulder-clays; Montreal; Quebec; Murray Bay; Rivière-du-Loup; Anticosti; New Richmond; Goose River; Chaudière Station; Duck Cove, St. John, N.B. (Matthew); Maine; Labrador; Greenland (Möller); also European Pleistocene.

Recent—Little Metis; Gaspé; Rivière-du-Loup; Arctic seas generally, and on the American coast south to Massachusetts.

This shell is extremely abundant in the Leda and boulder-clays, and often occurs in the clay with the valves attached. It is also of large size and in fine condition, especially at Rivière-du-Loup. It is *Tellina proxima*, Brown, *T. sabulosa*, Spengler, and *T. sordida* of Couthuoy. According to Hanley, the *T. lata* of Gmelin was founded on a figure of this shell. British Columbia.

Macoma inflata. Stimpson.

Fossil—Montreal; Rivière-du-Loup. Rare.

Recent—Murray Bay; and dredged in deeper parts of the gulf of St. Lawrence by Mr. Whiteaves.

I am not aware where this little shell has been described, nor what is its range. It seems identical with a specimen in Jeffrey's collection labelled *Tellina fragilis*, Leach, from Spitzbergen. The Pleistocene specimens are larger and better developed than the recent, except some dredged by Mr. Whiteaves on the north shore, and I would infer from this that the shell is Arctic.

Cyrtodaria siliqua. Daudin.

Fossil—Rivière-du-Loup; Labrador (Packard); Greenland (Möller). I have seen in the Post-pliocene of Canada, only an imperfect and decorticated specimen of the young shell from Rivière-du-Loup.

Recent—Little Metis; Cape Breton; Prince Edward Island; Gaspé; Gulf of St. Lawrence; and coasts of Nova Scotia and New England.

Mactra (Spisula) ovalis. Gould. *M. polynema.* Stimpson.

Fossil—Boulder clay; Cape Elizabeth, Maine.

Recent—Little Metis; Gaspé; Labrador (Packard); also coast of New England.

I found, many years ago, a few specimens of this shell at a cove where a number of species of marine shells occur in boulder-clay, and it was published in my list of shells from this place in my paper on the Post-pliocene of Labrador, Maine, &c. It is credited by Packard to "Zeeb's Cove," Cape Elizabeth, which may probably be the same place where I procured it. This species has not yet been found within the limits of Canada in the Pleistocene, though this and the related species or variety, *M. solidissima*, is found living in Labrador, and Matthew records it from upper Leda clay, St John. It has perhaps moved northward since the glacial period.

Mesodesma (Ceronia) deaurata. Turton.

Fossil—Matane River (Bell); Little Metis. I have not seen it in any other locality; and it occurs only on the lowest terrace, so that possibly it is modern.

Recent—Abundant at Tadoussac; Little Metis; and elsewhere in Gulf and River St. Lawrence; Labrador (Packard).

This must be a modern species on our coasts; but according to Wood it is found in the Red Crag of England.

Venericardia (Cardita) borealis. Conrad.

Fossil—Labrador (Packard).

Recent—Arctic seas to Long Island; Little Metis, and common throughout the Gulf of St. Lawrence. It would seem to have been much less generally distributed in the Pleistocene. Western America as far south as Catalina Island, P. P. C. British Columbia.

Astarte Laurentiana. Lyell.

Fossil—Leda clay, Montreal, abundant; Beauport and Rivière-du Loup, rare.

Recent—Greenland (Morch); Labrador (Packard); Murray Bay.

This shell may be a variety of the next species; but it is at least a very distinct varietal form. It is distinguished by its very fine and uniform concentric striation, passing to the ends of the valves and to the ventral margin. There are two varieties, a flatter, and a more tumid. I have the former from Greenland named by Morch *A. Banksii*, and the latter named *A. striata;* but they are different from shells indicated by these names in Gould and elsewhere. The only recent specimens that I have seen from the gulf of St. Lawrence, which can be referred to this species, are a few I dredged at Murray Bay. *A. Laurentiana* is very abundant at Montreal, but much more rare nearer the coast. It is evidently an arctic form. (See Figure, Plate I.)

Astarte (Nicania) Banksii. Leach.

Fossil—Leda clay, Rivière-du-Loup; Anticosti; Little Metis; Kamouraska, abundant; Quebec, not infrequent; Montreal, very rare; Labrador (Packard); St. John (Matthew); Portland, Maine, also Uddevalla, Clyde beds and Crag.

Recent—Abundant at Gaspé and elsewhere in Gulf of St. Lawrence, and also Arctic seas and coast of Nova Scotia.

This shell is that named *A. Banksii*, in Gould's last edition, also in Beechey's voyages. It is easily distinguished from the last species by its coarser striation, fading toward the ends and also toward the margin of the shell. It is, however, about the same size, but less delicate and symmetrical in form. It is the common small Astarte of the gulf of St. Lawrence, and also of the Post-pliocene of Rivière-du-Loup; but becomes very rare at Montreal, where it is replaced by *A. Laurentiana*. This species was named *A. compressa* in my former lists, and it is certainly very near to European specimens of that species, especially to the fossils from the Clyde beds and the Crag. (See Figure, Plate I.)

Astarte elliptica. Brown.

Fossil—Labrador; Saguenay; Portland, Maine.

Recent—Labrador; Murray Bay; Rivière-du-Loup; Little Metis; Kamouraska; Gaspé; coast of Nova Scotia, &c. Also Greenland; Norway (typical); Scotland.

Specimens from the Clyde beds are perfectly identical with ours. It is also found in the Post-pliocene of Norway and rarely in the Crag. It is a northern species, meeting on the American coast the closely allied forms *A. undata* and *A. lens*, into which, however, it does not seem to pass. The two latter species, being more southern forms, are not found in the Pleistocene. A small form of *A. crebricostata* (= lens) is very abundant in 200 fathoms gulf St. Lawrence, J. F. W.

A. Omalii of S. Wood from the Crag, is very near to this species, but is at least a distinct variety.

Astarte elliptica, Brown, has been shown by Sylvanus Hanley to be the *Venus compressa, Linn.* Hence it is the true *A. compressa.* J. F. W.

I regard this as *Astarte lactea* Brod and Sby.; and *A. semisulcata*, Leach, but as probably distinct, as Astartes go, from *A. borealis*, (= *A. arctica*). (See Figure, Plate I.)

Astarte arctica, Möller, (var. *lactea*.)

Fossil—Labrador (Packard); St. John (Matthew); Portland, Maine; also Greenland (Möller).

Recent—Gaspé; Little Metis; Rivière-du-Loup; also Arctic seas; Norway (typical).

This species has been found in the Pleistocene of Canada, only in Labrador and New Brunswick; and is rare in the gulf of St. Lawrence. It is our largest Astarte and I believe it to be identical with *A. lactea*, Brod. and Sow., and *A. semisulcata*, Gray. Fossil specimens from Portland are precisely similar to recent ones from Gaspé dredged by Mr. Whiteaves and referred by him to *A. lactea*. But he regards *A. borealis* as probably distinct. Specimens from Norway (*A. arctica*) and from Clyde beds (*A. borealis*) are smoother and less ribbed than ours. British Columbia. (See Figure, Plate I.)

Other species of Astarte.

At Murray bay, there occurs very rarely a transversely elongated and regularly striated Astarte with delicately wrinkled epidermis, which seems to be identical with *A. Richardsonii* from the Arctic seas as described but not as figured by Reeve; but *A. Richardsonii* is generally regarded as young *A. lactea*. A similar species or variety occurs, but very rarely, fossil at Rivière-du-Loup. Matthew mentions *A. compressa* from Pleistocene at St. John. *A. sulcata* (undata), *A. crebricostata* (= *A. lens*), *A. castanea*, and *A. quadrans* have not yet been found fossil, though the three former at least live in the gulf of St. Lawrence.*

Cardium pinnulatum. Conrad.

Fossil—Leda clay; Lawlor's Lake and St. John, N.B. (Matthew).
Recent—Gulf St. Lawrence, and coast of Nova Scotia and New England.

Cardium Islandicum. Linn.

Fossil—Rivière-du-Loup; Murray Bay; Saguenay; Little Metis; Vancouver Island (G. M. D.); Portland, Maine; Lawlor's Lake, N.B.; Greenland (Möller).
Recent—Rivière-du-Loup; Little Metis; from Greenland to New England.

Our fossil specimens are mostly small, and similar to the northern variety or sub-species named by Stimpson *C. Hayseii*, and which also occurs living as far south as Nova Scotia, and seems to be the *C. ciliatum* of Fabricius. Decorticated specimens are not distinguishable from *C. Dawsonii* of Stimpson, from the Pleistocene of Hudson's Bay; of which I have seen only specimens in this state.

Cardium (Serripes) Grœnlandicum. Chemnitz.

Fossil—Leda clay and boulder-clay, Quebec; Rivière-du-Loup; Murray Bay; Lawlor's Lake, &c., N.B. (Matthew); New Richmond; Restigouche; Bathurst (Paisley); Cape Elizabeth, Maine; Labrador (Packard); Greenland (Möller); Chaudière Station.
Recent—Little Metis; Gaspé; Gulf St. Lawrence, sometimes of large size; Arctic seas, and Greenland to Cape Cod.

This shell is somewhat rare and of small size in the Post-pliocene, and has not yet been found higher up the St. Lawrence than Quebec. Specimens of good size occur at Portland.

Kellia Suborbicularis. Mont.

Black Point, N.B. (Matthew).

---

* *Astarte quadrans*, Gould, has been dredged, living, off Esquimaux point, on the north shore of the St. Lawrence. *Astarte crenata*, Gray, and a small form of *A. crebricostate*, Forbes (= *A. lens*, Stimpson), have been dredged in 200 to 300 fathoms between Anticosti and the south shore. The form of *A. undata*, Gould, which comes closest to *A. sulcata*, is abundant, living, in Northumberland straits and between Cape Breton and Prince Edward Island. J. F. W. *A. castanea* occurs in Minas basin, Bay of Fundy.

Cryptodon Gouldii. Philippi.

Fossil—Montreal. Rare. Matthew records a species supposed to be distinct from *C. Gouldii* at St. John.

Recent—Murray Bay; Gaspé (Whiteaves); Little Metis; Kamouraska; Greenland to New England.

The European form *C. flexuosa (Axinus flexuosus)* is usually regarded as distinct, and is found as far north as Spitzbergen, and in the Crag, the Clyde beds, and the Norway Post-pliocene, and in British Columbia. Jeffreys, however, considers the difference merely varietal, and it certainly seems to diminish or disappear in the northern and glacial specimens.

According to Mr. Whiteaves this species has a great range in depth in the gulf of St. Lawrence, being found, living, from 20 to 300 fathoms.

Sphaerium?

Fossil—Pakenham Mills, with fresh-water bivalves and *Tellina Grœnlandica.* The specimens were too imperfect for certain determination.

Unio rectus. Lamarck.

Fossil—Clarenceville, Lake Champlain (Dickson), with *Mya arenaria, Tellina Grœnlandica,* &c.

Recent—River St. Lawrence.

Unio Cardium? Rafinesque.

Fossil—With the preceding. This and the preceding species were represented by large and thick shells better developed than those of the River St. Lawrence at present. It is probably the same with *U. ventricosus,* Barnes.

Unio ellipsis. Say.

Fossil—Toronto ; Interglacial Beds.

Recent—River St. Lawrence.

Mytilus edulis. Linn.

Fossil—Montreal; Acton ; Rivière-du-Loup; Quebec ; Chaudière Station ; Anticosti ; Labrador (Packard) ; Lawlor's lake, N.B. (Matthew); Greenland (Möller).

Recent—North Atlantic and Arctic seas generally ; British Columbia and North Pacific (= *trossulus,* Gould) ; as far south as Monterey.

The variety most commonly found in the Pleistocene is a small, oval, tumid form, allied to variety *elegans* of British writers. This variety still lives at Tadoussac, and is apparently characteristic of situations where the water is cold and exposed to intermixture of fresh water. The ordinary variety occurs at Portland, and also in some of the upper beds at Rivière-du-Loup. At Montreal only the small oval variety occurs. This variety is also found in the Clyde beds and in the Crag.

Modiola modiolus. Linn.

Fossil—Montreal, very rare.

Recent—Labrador to New England ; very common on the coasts of Nova Scotia and New England ; North Pacific ; found sparingly along the Vancouver and Californian coasts till it is replaced in the Gulf fauna by *M. capax*, Conrad.

This species becomes rare to the northward ; and this, as well as its being proper to rocky shores rather than to clays and sands, may account for its rarity in the Canadian Pleistocene. It is, however, common in the glacial beds of Europe.

Modiolaria nigra. Gray.

Fossil—Montreal ; Rivière-du-Loup (small variety *nexa ;* also large and fine) ; very large and well preserved in nodules at Kennebeck, Maine ; Labrador (Packard, if his *M. discrepans* as I suppose) ; Black Point, N.B. ? (Matthew). (See Plate V., Fig. 5.)

Recent—Gulf of St. Lawrence (Whiteaves) ; Little Metis ; Rivière-du-Loup. Very large and fine on coast of Nova Scotia (Willis) and as far north as Greenland (*M. discors*, L.) ; British Columbia.

Modiolaria corrugata. Stimpson.

Fossil—Rivière-du-Loup.

Recent—Murray Bay, Little Metis and Cacouna ; precisely similar to the shells from the Pleistocene. Also Greenland (Möller) ; Labrador (Packard) ; and south to Cape Cod.

Modiolaria discors.

Fossil—Beauport, of good size ; Greenland (Möller) ; Montreal (Mr. Kennedy).

Recent—Labrador to New England ; Little Metis ; Rivière-du-Loup ; British Columbia. Specimens from Gaspé are precisely similar to the fossil. This shell is no doubt identical with *M. lævigata* of Gray, and possibly with the *M. discrepans* of some other authors. It is, however, the same with that figured in Binney's Gould as *M. discors*.

Crenella glandula. Totten.

This species, which is at present quite common in the Gulf St. Lawrence, is indicated in my formerly published lists as a Montreal fossil; but I have mislaid the specimens, and cannot therefore now repeat the comparisons with the recent shells. It is probably *C. faba* of Fabricius.

According to Mr. Whiteaves this is quite distinct from *C. decussata,* Montagu, both being found living in Gaspé.

Nucula tenuis. Montagu.

Fossil—Leda clay, Montreal; Saco (var. *inflata*); Rivière-du-Loup, Bay de Chaleur (Matthew)? Green's Creek, Ottawa River.

Recent—North Shore; also Gulf St. Lawrence to Gaspé (Whiteaves); Little Metis (type and var. *inflata;* also European coasts and British Columbia.

N. expansa. Reeve.

Fossil—Leda clay and boulder-clay, Rivière-du-Loup; Saco; Westbrook, Duck Cove, St. John, N.B. (Matthew); Pakenham?

Recent—Labrador (Packard); Murray Bay; Little Metis; Arctic seas.

I doubt if this is not a large and well-developed northern form of *N. tenuis.*

N. antiqua. Mörch. From Leda clay of Maine; seems to be a variety of the last.

Leda pernula. Muller.

Fossil—Leda clay, Rivière-du-Loup; New Richmond; New Brunswick (Matthew); Portland; Saco; Lawlor's Lake, N.B. (Matthew).

Recent—Little Metis; Kamouraska; Arctic seas and south to New England.

This shell occurs very abundantly at Rivière-du-Loup; and the specimens found there show that no specific line can be drawn between the forms known as *pernula, buccata* (Steenst.), *tenuisulcata,* Gould, and *Jacksonii,* Gould. Slender and flattened varieties are *pernula* and *tenuisulcata,* shorter and more tumid forms are *buccata;* and specimens decorticated so as to show the origin of the hinge teeth are *Jacksoni.* Comparison of specimens from Greenland, Norway, Labrador, the Gulf St. Lawrence, and New England, confirms this conclusion.

Leda minuta. Fabricius.

Fossil—Leda clay, Montreal; Rivière-du-Loup; St. John, &c. (Matthew); Greenland (Müller); Labrador (Packard).

Recent—Little Metis; Kamouraska; also British Columbia; Arctic seas, Gulf St. Lawrence; coast of Nova Scotia.

The fossil specimens occur abundantly with the last species at Rivière-du-Loup, and are quite similar to those dredged at Murray Bay. This was called *L. caudata* in my former lists.

Leda pygmæa. Munster.

Fossil—Leda clay, Green's Creek, Ottawa; Saco, Maine; also English Crag and other Glacial beds.

Recent—North European seas; but not yet recognized on the American coast. According to Mr. Jeffreys and Dr. Carpenter, our drift-shells are referable to the variety or species *Yoldia abyssicola* of Torell.

Yoldia (Portlandia) arctica. Gray. (= *Leda glacialis,* ante.)

Fossil—Leda clay and Boulder clay, Montreal; Quebec; Ottawa River; Rivière-du-Loup; St. John, N.B., &c.; Portland and Saco, Maine; also in Pleistocene of Norway (Sars), and of Scotland (Crosskey).

Recent—Arctic seas.

This shell is most abundant, and generally diffused in the Leda clay; and the variety ordinarily found at Montreal and Rivière-du-Loup is precisely identical with the ordinary Arctic form. A long variety, called *L. intermedia* by Sars, is also found at Montreal, though rarely. A short variety, common in the Pleistocene at Murray Bay, is similar to the *L. siliqua* of Reeve from the Arctic seas; and young and depauperated varieties resemble *L. sulcifera* of the same author. The abundant material from the Pleistocene shows that these are all varietal forms. (Plate V., Fig. 4.)

This shell is *Yoldia arctica* of Sars, but not of Müller and Morch. It is *Y. truncata* of Brown. It is *Portlandia glacialis* of Gray, and *Leda Portlandica* of Hitchcock.

Yoldia lucida, Lovén (which is abundant living in the deeper parts of the Gulf of St. Lawrence) resembles the young form of this species, but the two are probably quite distinct. J. F. W.

Yoldia limatula. Say.

Fossil—Leda clay, Rivière-du-Loup. *Yoldia sapotilla* is recorded by Matthew from Black Point, N.B.

Recent—Little Metis; Gulf St. Lawrence to Long Island.

17

This shell has been found as yet only at Rivière-du-Loup, where the specimens however are as good as those now living in the Gulf. It will be observed, however, that though they have the number of teeth of *Y. limatula*, they approach in form to the allied species or variety *Y. sapotilla*, a shell which occurs in Greenland and thence to New England, and which I strongly suspect is merely a short variety bearing a similar relation to *Y. limatula* to that which *Mya Uddevallensis* bears to the ordinary *M. truncata* ; but Jeffreys considers it distinct. *Y. sapotilla* is, I may mention, the *Y. arctica* of Morch, as proved by a specimen from his collection now in my possession.

Yoldia myalis. Couthuoy.

Fossil—Labrador (Packard).

Recent—Rivière-du-Loup ; Little Metis ; Kamouraska ; Gaspé (Whiteaves) to south of Cape Cod. This shell is supposed to be identical with *hyperborea*, Lovén, from Spitzbergen.

Leda fossa. Baird.

Fossil—Vancouver Island (Dr. G. M. Dawson).

Pecten Grœnlandicus. Chemnitz.

Fossil—Leda clay, Portland and Saco, Maine ; not yet found in Canada.

Recent—Gulf St. Lawrence (Whiteaves), in deep water 200 to 300 fathoms.

This species is found in the Clyde beds and in Greenland ; and if, as Jeffreys supposes, identical with *P. similis* (Laskey), it is a shell of very wide distribution in the Atlantic, as well as in geological time. Though not yet found in Canada as a Pleistocene fossil, its occurrence as a fossil in Maine and recent in the Gulf St. Lawrence, renders it probable that it may yet occur in our Leda clays.

Pecten tenuicostatus. Mighels. (= *P. Mogellanicus.* Lamarck.)

Fossil—Leda clay, St. John, N.B. (Matthew).

Recent—Labrador to Cape Cod.

This shell has not yet been found in the Pleistocene of the St. Lawrence valley; but since, according to Packard, it is common in Labrador, there is nothing remarkable in its occurrence at St. John.

Pecten Islandicus. Chemnitz.

Fossil—Rivière-du-Loup; Quebec ; Anticosti; Labrador (Packard); St. John, N.B. (Matthew); Portland, Maine ; Greenland (Möller); also Crag, Clyde beds, and Post-pliocene of Norway.

Recent—Little Metis ; Murray Bay; Gaspé ; Gulf St. Lawrence, and from Greenland to Connecticut.

This is a shell which is very durable, and retains even its colour when imbedded in the clays. In this it excels the Tellinas, Astartes, Saxicava and Ledas ; though these in turn are always much better preserved than the Mytili and Modiolæ.

Ostrea Virginiana. Lister.

I have picked up a loose specimen at Saco which has the appearance of being a fossil specimen from the Leda clay, and Mr. Paisley has sent me specimens from the Bay des Chaleurs, which are said to have come from Pleistocene beds 16 feet from the surface.*

Class III.—Gasteropoda.

Philine lineolata. Couthuoy.

Fossil†—Montreal, rare.

Recent—Gaspé ; Grand Manan (Stimpson) ; Nova Scotia (Willis). It is *Philine lima,* Brown, according to Jeffreys.

Cylichna alba. Brown.

Fossil—Montreal ; Rivière-du-Loup ; also in the Clyde beds. (Plate VI., Fig. 11.)

Recent—Gaspé ; Labrador (Packard) ; Gulf St. Lawrence, common (Whiteaves) ; Arctic seas generally. Same or similar on West Coast at Sitka (P. P. C.)

Cylichna oryza. Totten.

Fossil—Montreal.

Recent—Coast of New England. Jeffreys regards it as *B. utriculus.* Povichi.

Cylichna nucleola. Reeve.

Fossil—Montreal ; rare, and perhaps doubtful.

Recent—Arctic seas.

---

* At present, however, its occurrence as a Pleistocene species must be regarded as doubtful, more especially as in modern times it does not occur in the colder parts of the Gulf of St. Lawrence.

† Except when otherwise stated, all the Gasteropods are found in the Leda clay, or at its junction with the Saxicava sand.

FOSSILS.—PLATE VI.

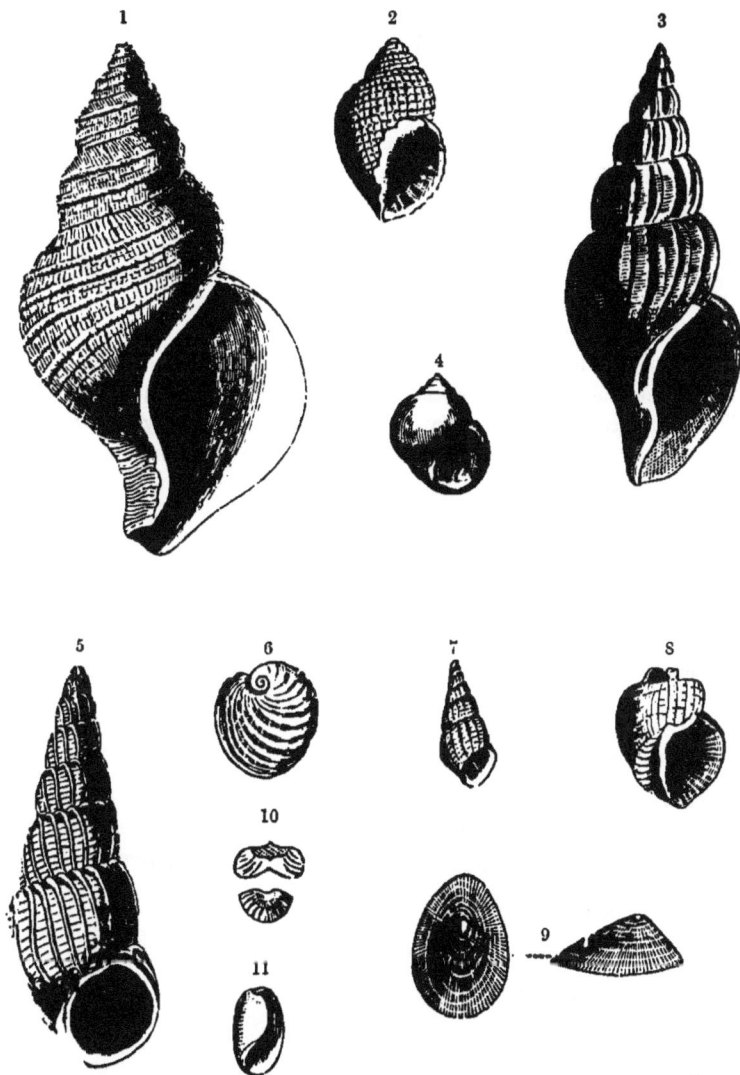

Some Characteristic Gastropods.—1, *Neptunea despecta*; 2, *Admete viridula*; 3, *Tritono-fusus Kroyeri*; 4, *Natica clausa*; 5, *Scalaria Grœnlandica*; 6, *Velutina zonata*; 7, *Acirsa Eschrichtii*; 8, *Trichotropis Arctica*; 9, *Lepeta cacca*; 10, *Amicula Emersonii*; 11, *Cylichna alba?*

Cylichna occulta. Mighels and Adams.

>Fossil—Montreal ; Murray Bay ; Maine.
>Recent—Greenland to New England.

Cylichna striata. Brown.

>Fossil—Rivière-du-Loup and Clyde beds.
>Recent—Arctic seas. Jeffreys thinks this the same with the preceding species.

Haminea solitaria. Say.

>Fossil—Montreal ; rather common.
>Recent—New England and northward.

If this species is rightly determined, it furnishes a curious instance of a somewhat southern species occurring in the drift of Montreal. The *Haminea*, however, can scarcely be identified by weathered or fossil specimens, so that this may possibly be a northern form distinct from *solitaria.*

Diaphana debilis. Gould.

>Fossil—Montreal.
>Recent—Gulf St. Lawrence (Whiteaves) ; Greenland to New England.

Jeffreys considers it the same with *B. hyalina*, Turton. If so, it is a shell of the Clyde beds and of the Arctic seas generally.

Utriculus pertenuis. Mighels.

>Fossil—Montreal.
>Recent—Labrador (Whiteaves) ; Gulf St. Lawrence, and south to Cape Cod. According to Jeffreys it is *U. turritus*, Möller, Greenland.

Patula striatella. Anthony.

>Fossil—Pakenham, Saxicava sand.

Lymnea umbrosa. Say.
>Fossil—Montreal.

Lymnea caperata. Say.
>Fossil—Montreal.

Limnæa palustris. Muller. (= *L. elodes*, Say.)
>Fossil—Pakenham Mills, Saxicava sand.

Planorbis bicarinatus.   Say.

Fossil—Pakenham Mills, Saxicava sand.

Planorbis trivolvis.   Say.

Fossil—Pakenham Mills, Saxicava sand.

Planorbis parvus.   Say.

Fossil—Pakenham Mills, Saxicava sand.

All of the above pulmonates are modern Canadian species, and seem to have been drifted by some fresh-water stream into the sea of the Saxicava sand and Leda clay.

Siphono-dentalium vitreum.   Sars.

Fossil—Leda clay, Murray Bay; also Norway (Sars).

Recent—Gulf of St. Lawrence (Whiteaves); coast of Norway (Sars). It is a rare deep-water shell.

Amicula (Stimpsonella) Emersonii.   Couthuoy.

Fossil—Montreal.   (Plate I., Fig. 10.)

Recent—Murray Bay; Little Metis; Rivière-du-Loup; Halifax; coast of New England.

My specimens are merely detached valves.   They indicate an animal quite similar to specimens from Halifax referred to this species, but differ slightly from specimens from Murray Bay.   Dr. Carpenter has labelled the drift form var. "*altior.*"   The differences among the recent specimens, as well as the fossil valves, are discussed in the "Contributions to a Monograph of the Chitonidæ," prepared by Dr. Dall partly from Dr. Carpenter's notes, and printed by the Smithsonian Institution.

Puncturella (Cemoria) noachina.   Linn.

Fossil—Quebec; Rivière-du-Loup; Clyde beds.

Recent—Little Metis; Gaspé; Murray Bay; Labrador; Gulf St. Lawrence generally; and throughout the Arctic seas and North Atlantic.

Acmæa testudinalis.   Möller.

Fossil—Labrador.

Recent—Tadousac; Little Metis; Rivière-du-Loup; Murray Bay; Gaspé; Pictou; Gulf St. Lawrence generally; and throughout the Arctic seas and North Atlantic.

My only fossil specimen, obtained from Dr. Packard, is of the small, elevated and depauperated variety so common at Murray Bay and the north shore of the Gulf. It is curious that this common modern species is so very rare in the Pleistocene.

Lepeta cæca. Möller.

Fossil—Montreal ; Rivière-du-Loup ; Quebec ; Labrador ; European Post-pliocene. (Plate VI., Fig. 9.)

Recent—Gaspé ; Labrador ; Arctic seas generally; and coast of New England rarely.

This shell is not at all rare, living at Gaspé, and fossil at Rivière-du-Loup. Carpenter remarks that some of my Montreal specimens have the characters of variety *striata* of Middendorff from Siberia.

Capulus Ungaricus. Lin. (*commodus ?* Middendorff.)

Fossil—Point Levi, near Quebec. One specimen only, found by Mr. Gunn and communicated by Dr. W. J. Anderson. Another and larger collected at Montreal (Prof. Kennedy) ; Scotland (Jeffreys). (Plate I., Fig. 14.)

Recent—Spitzbergen; Greenland ; Norway. I have not found this shell recent in the Gulf of St. Lawrence.

This species is fossil at Uddevalla, and is supposed to be the same with *C. fallax* and *C. obliquatus* of Wood from the English Crag. If a form of *Capulus Ungaricus*, it has been repeatedly dredged alive off the New England coast, in deep water, by Verrill.

Crepidula fornicata, L.

Baron de Geer, in his visit to Canada, 1891, was so fortunate as to find at the Mile-End Quarries, in the Saxicava sand, a small specimen of this shell, the first hitherto found in Canadian Pleistocene.

Margarita helicina. Fabricius.

Fossil—Montreal ; Murray Bay.

Young specimens resemble *M. acuminata* of Mighels. Broad specimens resemble *M. Campanulata*, Morse.

Recent—Arctic seas ; Gulf of St. Lawrence ; and coast of New England. Also British Columbia. It is *M. Arctica*, Leach.

Margarita argentata. Gould.

Fossil—Montreal. Rare.

Recent—Labrador and Gulf St. Lawrence (Whiteaves) ; Murray Bay ; Gaspé ; Rivière-du-Loup ; Little Metis ; coast of New England and Nova Scotia ? Possibly the same with *M. glauca*, Möller, from Greenland.

Margarita cinerea. Couthuoy.

Fossil—Rivière-du-Loup ; Portland.

Recent—Gaspé ; Labrador ; Rivière-du-Loup ; Little Metis ;
Murray Bay; Greenland to New England ; var. *striata*, Dall, Sitka.

Cyclostrema (Mölleria) costulata. Möller.

Fossil—Montreal ; Clyde beds ; Uddevalla.

Recent—Gaspé ; Arctic seas to New England.

Cyclostrema Cutleriana. Clark.

Fossil—Montreal. Rare.

This is an Arctic and British shell, as yet recognized only at
Montreal.

Turritella erosa. Couthuoy

Fossil—Labrador ; Rivière-du-Loup ; Montreal ?

Recent—Little Metis ; Murray Bay ; Nova Scotia ; Greenland to
New England.

Turritella reticulata. Mighels.

Fossil—Labrador (Packard).

Recent—Gaspé ; Labrador to Gulf St. Lawrence ; also fishing banks,
Nova Scotia (Willis) ; British Columbia.

My specimens received from Dr. Packard are marked *T. costulata*,
but seem rather to be the above species.

Turritella acicula. Stimpson.

Fossil—Rivière-du-Loup ; Labrador (Packard).

Recent—Murray Bay ; coast of New England.

There may be some reason to doubt whether this is not a variety of
T. erosa. It is quite possible that the above species should be regarded
as *Mesaliæ*.

Campeloma decisum. Say. (= *Paludina decisa*, Say.)

Fossil—Pakenham Mills, Saxicava sand.

Recent—Eastern America generally.

Valvata tricarinata. Say.

Fossil—Pakenham Mills, with the preceding.

Recent—Eastern America generally.

Jeffreys regards this as a variety of *V. piscinalis*.

Amnicola limosa. Say.

Fossil—Pakenham Mills, with the preceding.
Recent—Hudson's Bay to Virginia.
This was *A. porata* of the previous lists.

Littorina rudis. Martin.

Fossil—Rivière-du-Loup ; also Clyde beds and Uddevalla.
Recent—Arctic seas to New England and European coasts.
L. tenebrosa, which may be regarded as a variety, is also found at Rivière-du-Loup, Little Metis, and Kamouraska.

Rissoa castanea. Möller.

Fossil--Montreal.
Recent—Gaspé ; Labrador, Trinity Bay (Whiteaves).

Rissoa exarata. Stimpson.

Fossil—Montreal.
Recent—New England.

Cingula Jan Meyeni. Friele. (*Rissoa scrobiculata* of former lists.)

Fossil—Montreal.
Recent—Greenland ; Gulf St. Lawrence, 200 to 300 fathoms, large ; and small, Gaspé, 30 fathoms (Whiteaves).

Bela harpularia. Couthuoy.

Fossil—Montreal ; Quebec ; Murray Bay ; River Charles, &c., New Brunswick (Matthew) ; Rivière-du-Loup ; Labrador (large specimens).
Recent—Gulf St. Lawrence ; very fine at Murray Bay, and similar to large specimens from Rivière-du-Loup ; Rivière-du-Loup ; Little Metis ; Kamouraska ; coast of New England. It is *B. Woodiana,* Möller. (J. F. W.)

Bela elegans. Möller.

Fossil—Montreal ; Murray Bay.
Recent—Greenland and Norway ; closely allied to next species.

Bela pyramidalis. Ström.

Fossil—Montreal ; also Crag, Clyde beds and Uddevalla.
Recent—Labrador (Packard) ; Gulf St. Lawrence (Whiteaves) ; Murray Bay ; Rivière-du-Loup ; Kamouraska, and south to Cape Cod ; Arctic seas generally. It is the *B. pleurotomaria* of Couthouy, and

B. Vahlii of Beck. According to Dr. Carpenter the species is an uncertain one, having apparently varieties connecting it with *B. harpularia, B. bicarinata,* and *B. violacea.*

Bela turricula. Montagu.

Fossil—Montreal ; Rivière-du-Loup; Labrador ; also Red Crag and Uddevalla (Jeffreys).

Recent—Little Metis ; Gulf of St. Lawrence and coast of Nova Scotia and New England.

I include under this name *B. nobilis* of Müller ; *B. Americana,* Packard ; *B. scalaris,* Müller ; *B. exarata,* Muller, Morch ; and *B. angulata,* Reeve. The var. *nobilis* is found at Montreal and Gaspé ; also young shells not distinguishable from *exarata.* Var. *scalaris,* occurs at Rivière-du-Loup and Labrador. This shell is a widely-diffused and somewhat variable northern. species. Mr. Whiteaves, however, regards *B. nobilis, B. exarata,* and *B. scalaris* as distinct.

Bela Trevelliana. Turton.

Fossil—Rivière-du-Loup ; Labrador ; also Clyde beds and Norway (Jeffreys).

Recent—Little Metis ; Rivière-du-Loup; Murray Bay; Arctic seas, and Greenland to Massachusetts. It is probably *B. decussata* of Couthuoy. *B. excurvata* (Carpenter), from Puget Sound, may prove another variety. British Columbia, according to Whiteaves.*

Bela violacea. Mighels and Adams.

Fossil—Montreal.

Recent—Little Metis ; Rivière-du-Loup ; British Columbia ; Murray Bay ; Labrador (Packard); Gaspé (Whiteaves); Fishing banks Nova Scotia (Willis); Massachusetts (Stimpson). Possibly *B. bicarinata* of Couthuoy.

Bela cancellata. Mighels and Adams.

Fossil—Little Metis ; Murray Bay ; Labrador (Packard); Casco Bay (Gould). This shell may be *B. impressa,* Beck. In any case the fossils are identical with the modern Murray Bay specimens. It also occurs living in Gaspé Bay (Whiteaves).

---

* Whiteaves doubts whether the true *Bela turricula* or the *B. Trevelyana* have been found in the St. Lawrence.

Natica clausa. Brod. and Sowerby.

Fossil—Montreal ; Quebec ; Rivière-du-Loup ; St. John, &c., N.B. (Matthew) ; Labrador ; Anticosti ; Chaudière Station ; Portland, Maine ; Vancouver Island. (G. M. D.) (Plate VI., Fig. 4.)

Recent—Greenland to Cape Cod ; Little Metis ; Gaspé ; Tadousac ; Kamouraska ; also British Columbia.

Common and extensively distributed in the Post-pliocene of Europe, from Norway to Sicily, and found at an elevation of 1330 to 1360 feet in Moel Tryfaen, Wales. (Darbyshire). It has been identified with *N. affinis*, Sm., but Sars and Verrell consider this quite different.

Lunatia heros. Say.

Fossil—Beauport, a single specimen only, and this of small size ; River Charles, N.B. (Matthew); New Richmond; also a depauperated variety ; Benjamin River, supposed by Matthew to be possibly a distinct species or variety. Var. *Chalmersi* (Matthew).

Recent—Labrador and southward.

This species is as old as the Miocene Tertiary ; and in the Pleistocene, Canada was probably its extreme northern limit.

Lunatia Grœnlandica. Beck.

Fossil—Montreal ; Quebec ; Rivière-du-Loup ; Anticosti ; Maine ; also Pleistocene of England, Scotland, and Norway.

Recent—Little Metis ; Gaspé ; Rivière-du-Loup ; Arctic seas generally ; and extending to Britain and New England.

L. pallida is the representative of this species on the west coast of America.

Choristes elegans. Cpr.

Fossil—Saxicava sand, Montreal. Rare. (Plate I., Fig. 13.)

This shell was identified in my former papers with *Natica helicoides;* but it is now found to be quite distinct, and Dr. P. P. Carpenter describes it as a new species and genus as follows :

Genus CHORISTES.

Testa helicoidea, tenuis ; epidermide induta ; anfractus disjuncti ; labrum postice angulatum, antice haud emarginatum; labium planatum ; columella simplex. Animal ignotum.

Choristes elegans, n. s.

Ch. t. satis elevatâ, tenui, nitente ; epidermide fulvâ, tenui, lævi, extus et intus omnino appressâ; anfr. iii. + ...?, vertice nucleoso decollato, spiraliter obsoletius striatis ; lineis incrementi tenuissimis ;

spirâ superne planatâ, suturis maxime impressis, basi tumente ; umbilico intus majore, extus modico ; aperturâ sublunatâ, postice ad angulum circ. 30° inclinatà, antice late rotundatâ ; labro acuto, postice planato ; labio acuto, planato, haud reflexo ; columellâ postice regulariter arcuatâ, neque emarginatâ, nec angulatâ, nec insculptâ.

Long. (apice decollato) ·82, *long. spir.* ·32, *lat.* ·76 poll. *Div.* 90°.

Hab. Montreal, in strato glaciali, fossilis, rarissime reperta. Mus. Dawson, McGill Coll., Nat. Hist. Soc.

Dr. Carpenter adds the following remarks :

While almost all the other drift fossils are of species still living in the neighbouring seas, this is not known, even generically, to be at present in existence. It is hard to pronounce satisfactorily on its relationships. In its thin, coated shell it resembles Velutina ; the striae and loose whirls recall Naticina ; the straight pillar lip reminds us of Fossarus ; while the umbilicus and rounded base, with entire mouth, best accord with the Natica group. With Trichotropis and its congeners I can see no resemblance. One remarkable feature in all the specimens is the decollation of the upper whirls, seen even in a nearly perfect young specimen, ·2 across ; other young specimens, even smaller, have only one whirl and a half remaining. The broken portion is filled up not so much by a septum as by a solid thickening. The separation of the whirls is complete from the beginning ; and although, in the parietal portion, they are closely appressed, the smooth and somewhat glossy epidermis is distinctly seen between. The fracture of the mouth in most of the specimens, enables this feature to be distinctly observed ; and would also reveal the "internal groove" and columellar callosity ascribed to Torellia, did any such exist.

The straightening of the inner lip, at an angle of 30° from the axis, makes the umbilicus by no means large (for a Naticoid shell) when viewed from the base in the line of the pillar ; but the same cause enlarges it within, recalling the adult appearance of Amphithalamus. The flattening of the upper portion of the whirls gives the shell somewhat of an Ianthinoid aspect.

While the analogies of the shell point in so many different directions, it is impossible to assign it even to its family group. It is to be hoped, however, that the dredge will yet reveal its existence in a living state.

The above species may be supposed to resemble *Torellia vestita,* Jeffreys, from Norway. Our specimens differ however in form, as above noted, and also in the absence of the tooth in the inner lip, and in the smooth epidermis.

The shell in question presents the very unusual character of having the whirls appressed, yet quite disconnected ; the smooth epidermis lining the umbilical chambers, and conspicuously preserved, even in these fossil specimens, between the closest parts of the parietal region. In this respect it bears the same relation to Torellia as does Latiaxis to Rapana, Separatista to Rhizochilus, or Zanclea to Torinia. It presents a rude resemblance to *Separatista Chemnitzii* (Add. Gen. pl. xiv. f. 6), or still more to *S. Blainvilleana* (Chênu Man. p. 172, § 853), but without the grooved pillar, or the keels of the latter species.

As to the " blunt tubercle " or " callous protuberance " of Torellia, described by Mr. Jeffreys, but scarcely to be traced in Mr. Sowerby's figure, it certainly does not exist in our fossils. It is not always a character of importance, as may be seen by comparing *Purpura columellaris* with *P. patula*, *Cuma tectum* with the remaining species of the genus, or the gradual transition from Isapis to Fossarus. The Naticidæ are often very irregular in the callous region of the pillar, even in the same species.

[The late J. Gwyn Jeffreys regarded it as congeneric with his *Torellia vestita* from the North Atlantic, but specifically distinct. Verrill has, since the discovery of the species in the Pleistocene, dredged specimens in deep water off the New England coast.]

Velutina zonata. Gould. (Plate VI., Fig. 6.)

Fossil—Montreal ; Beauport.
Recent—Arctic seas to Massachusetts; Little Metis; Kamouraska ; Nova Scotia.
According to Jeffreys, this shell is the same with *V. undata*, Smith, from the Clyde beds, and is found in the Crag and in the Post-pliocene of Uddevalla.

Scalaria Grœnlandica. Perry.

Fossil—Rivière-du-Loup ; Quebec ; Saco ; also Scottish Post-pliocene and English Red Crag, under same varietal forms as in Canada. (Plate VI., Fig. 5.)
Recent—Rivière-du-Loup; Nova Scotia ; Kamouraska ; Arctic seas, and American coast, as far south as Massachusetts.
The specimens from Rivière-du-Loup are very large, one being nearly two inches long ; and, as Dr. Beck has remarked, the varices of some of the specimens are more slender and lamellar than in recent specimens, others, however, are similar to the more common recent variety.

Acirsa Eschrichtii. Holboll.

Fossil—Quebec; Rivière-du-Loup; Montreal; most abundant at Rivière-du-Loup. (Plate VI., Fig. 7.)

Recent—Murray Bay; Little Metis; Greenland; also Eastport (Verrill).

This shell was named in former papers *Menestho albula,* the eroded specimens found being referred to that species. It has, however, been correctly described by Dr. Beck in Lyell's paper on Beauport, and named *Scalaria borealis.* It is not this species of Gould, however.

Trichotropis borealis. Brod. and Sow.

Fossil—Montreal; Rivière-du-Loup; Chaudière Station; Labrador, &c. Very abundant at Montreal.

Recent—Labrador; Little Metis; Murray Bay; Gaspé; Arctic seas; and as far south as Massachusetts; Alaska (Dall).

Trichotropis arctica? Middendorff.

Fossil—Montreal. Very rare. (Plate VI., Fig. 8.)

A single imperfect specimen represents this species, which is recent at Alaska (Dall) and Behring's Straits. The identification is perhaps doubtful.

The figure given by Reeve of *T. Kenseri* of Phillippi from Spitzbergen, resembles our shell, except in the small number of revolving bands.

Admete viridula. Fabricius.

Fossil—Montreal; Chaudière Station.

Recent—Labrador (Packard); Little Metis; Rivière-du-Loup; Murray Bay; Gaspé (Whiteaves); also Greenland, Labrador and British Columbia. It is the *Tritonium viridulum* of Fabricius, and is a rare shell in the Canadian Pleistocene, and in the Gulf of St. Lawrence.

Aporrhais occidentalis. Beck.

Fossil—Labrador (Bayfield); also Packard.

Recent—Little Metis; Gaspé; Labrador to Massachusetts.

It is remarkable that this species, which is found living from Labrador to Cape Cod, is so rare in the Pleistocene.

Ptychatractus ligatus (= *Fasciolaria ligata,* Mighels).

Fossil—Montreal; very rare.

Recent—Murray Bay; Mingan (Foote); Gaspé (Whiteaves); Nova Scotia (Willis); rare in all these localities.

A single mutilated specimen alone, as yet, represents this species in my Pleistocene collections.

Astyris Holbollii. Möller.

Fossil—Rivière-du-Loup ; also glacial beds Britain (Jeffreys).

Recent—Little Metis ; Kamouraska ; Gaspé ; Murray Bay ; Labrador (Whiteaves). If identical, as I suppose, with *Columbella rosacea* (Gould), it extends south to New England, and Gould's name has priority.

Buccinum undatum. Linn.

> var. *undulatum.* Möller.
>
> var. *Labradoricum.* Reeve.

Fossil—Saxicava sand and Leda clay, Rivière-du-Loup; Anticosti ; Labrador ; Duck Cove, St. John, N.B., (Matthew) ; Maine (Packard).

Recent—Little Metis ; Gaspé ; Kamouraska, &c. ; Gulf St. Lawrence ; south Greenland to Nantucket. (See figure, Plate I.)

I cannot satisfy myself that there is any good specific distinction between this shell and *B. undatum* of the European seas and glacial beds. It varies very much in size, in slenderness, in the fineness of the spiral striation, in the development of the ribs, in the extension of the mouth, and in the thickness of the shell. The coarser forms are *B. Labradoricum*, which passes into the ordinary *undatum*. Medium varieties are *B. undulatum* and smooth varieties pass into *B. cyaneum* and *B. Tottenii*, which last is the *ciliatum* of Gould.

Buccinum Tottenii. Stimpson.

Fossil—Rivière-du-Loup, Saxicava sand and Leda clay.

Recent—Little Metis ; Murray Bay and Tadoussac ; also Newfoundland Banks. It has some resemblance to *B. Humphreysianum*, Bennet, but is specifically distinct. It is the *B. ciliatum* of Gould, but has no connection with the *ciliatum* of Fabricius, except a slight resemblance to the smoother forms of the latter. It is remarkable for its very regular spiral lines, absence of folds and convex whorls.

Buccinum cyaneum. Bruguière.

Fossil—Rivière-du-Loup, abundant.

Recent—Murray Bay and Tadoussac ; Little Metis ; deeper parts of Gulf St. Lawrence (Whiteaves) ; Arctic seas.

This species or varietal form is well represented in the figure, which is taken from a large Rivière-du-Loup specimen. Being on the one hand very near to if not identical with the smooth varieties of *B.*

undulatum, and on the other resembling *B. Grœnlandicum*, it has received many names. It is believed to be *B. boreale* of Leach, and *Grœnlandicum* of Morch. It is a very characteristic northern form. (See figure, Plate I.)

Buccinum Grœnlandicum. Chemnitz.

Fossil—Leda clay and boulder clay, Montreal; St. Nicholas; Rivière-du-Loup; Tattagouche River (Paisley).

Recent—Greenland; Alaska? (Dall); Little Metis; Murray Bay. Specimens from Morch are identical with our fossils. This species is probably the *B. undatum* of Fabricius. It is allied to *B. cyaneum*, and may possibly pass into it. It may be *B. angulosum* (Gray). (See figure, Plate I.)

Buccinum tenue. Gray.

Fossil—Rivière-du-Loup, not uncommon; St. John, &c., N.B. (Matthew); Greenland (Hayes); Labrador (Packard).

Recent—Little Metis; Murray Bay; Gaspé; Labrador (Packard); Alaska (Dall); Arctic seas generally. A common arctic species, and now living in the Gulf, though much more plentiful in the Pleistocene beds. (See figure, Plate I.)

Buccinum ciliatum. Fabricius.

Fossil—Montreal; Rivière-du-Loup.

Recent—Murray Bay; Little Metis, Rivière-du-Loup; Greenland (Fabricius); Nova Scotia (Willis); Alaska (Dall).

This is the original *B. ciliatum* of Fabricius, and has been recognized as such by Dr. Stimpson. It is easily distinguished by its narrow Nassa-like mouth, armed with a tooth on the front of the pillar lip. It varies much in colour, especially in the longitudinal ribs. The variety found at Montreal is only slightly ribbed. That at Rivière-du-Loup is more distinctly ribbed, thus resembling the recent specimens from Murray Bay. It is quite distinct from *B. ciliatum* (Gould), which is very near the smoother varieties of *B. undulatum*. As it is a rare and little known shell, I have figured two extreme varieties, a fossil specimen from Montreal and a recent from Murray Bay. I submitted specimens of this shell to the late Mr. J. Gwyn Jeffreys in 1876, and after comparison with the type in Copenhagen he agrees with me in referring them to Fabricius' species. He says it is the species figured by Reeve as *B. Moelleri*.

Buccinum glaciale. Linn.

Fossil—Rivière-du-Loup; Montreal; Anticosti; Labrador (Packard); Black Point, N.B. (Matthew).

Recent—Murray Bay; Little Metis; Alaska (Dall); Greenland, and Arctic seas generally.

This shell has the aperture somewhat like that of *ciliatum*, and a very peculiar sculpture of spiral striae with intervening bands marked with finer striae. It has also a carina angulating the body whorl, and sometimes more than one. In the latter case it passes into *B. Grœn-landicum*, Hancock (not Chemnitz) or *B. Hancocki* Morch. The ordinary variety is most common in the modern Gulf, the latter in the arctic seas and in the Pleistocene. This shell, usually much decorticated, is the most common Buccinum in the Pleistocene of Montreal. (See figures, plate I.)

Buccinum plectrum. Stimpson.

Fossil—Rivière-du-Loup; rare.

Recent—Murray Bay; Portland, Maine (Stimpson); Behring's Straits (Stimpson); Alaska (Dall).

This may be a variety of the preceding species, but can be distinguished from it and grows to a larger size. It has the sculpture of *B. glaciale* with the aperture of *B. undulatum*. Recent and fossil specimens are quite similar.

The northern *Buccina* are involved in so much confusion that I have made them a subject of special study, and have sedulously collected all the forms recent and fossil. I have been very much aided in this by the abundance of specimens of the more arctic forms at Rivière-du-Loup, and the occurrence of most of them recent at Murray Bay and Tadoussac, and I feel confident that the names given in this list represent forms actually occurring as distinct in nature, though some of them may not be distinct specific types. I believe, however, that *B. ciliatum*, *B. glaciale*, *B. undulatum*, *B. tenue* and *B. Grœnlandicum*, are probably entitled to this rank. The others appear to me, on comparison of large numbers of specimens, to graduate into one or other of the above forms.

I have given in the engraved plate representatives of the more critical forms, which will enable them to be recognized.

In the drift the Buccina often part with their outer coat of prismatic shell, and in this decorticated state are very difficult to determine.

18

Tritonofusus Kroyeri. Möller.

Fossil—Rivière-du-Loup ; Labrador (Packard) ; New Richmond ; River Charles, N.B. (Matthew).

Recent—Little Metis ; Gulf St. Lawrence and Arctic seas. First recognized as this species by Mr. Whiteaves. Specimens from Spitzbergen in Mr. McAndrew's collection are perfectly similar to ours. Packard found it not uncommon at Labrador, but it seems rare in other parts of the Gulf of St. Lawrence, In some previous lists it has appeared as *B. cretaceum* (Reeve), which seems to be an error. Alaska (Dall). (Plate VI., fig. 3.)

Sipho Spitzbergensis. Reeve.*

Fossil—Montreal (small and rare).

Recent—Little Metis ; Murray Bay to Gaspé ; also Spitzbergen, and probably Sea of Okotsk ; N.W. coast (Dall).

Only one specimen occurred at Montreal, and was an unknown form until I fortunately dredged a few specimens at Murray Bay. It is a beautiful species, evidently quite distinct from *S. Islandicus.* From Middendorff's description and figure, I think it not improbable that it may be the same with his *Tritonium Schantaricum*, from the sea of Okotsk. I was not aware that it had been found on our coast, except at Murray bay, until these sheets were going through the press. Young specimens are remarkably like in form and sculpture to *Fasciolaria ligata*, which is found with it at Murray bay. Reeve's figure in Belcher's "Last of the Arctic Voyages," well represents our specimens, though perhaps a very little coarser in sculpture.

Neptunea despecta, L.

Fossil—Montreal ; Quebec ; Rivière-du-Loup ; Murray Bay ; New Richmond, River Charles, &c., N.B. (Matthew) ; Labrador (Packard).

Recent—Little Metis, large specimens ; Gaspé Bay, large specimens (Whiteaves) ; Labrador (Packard).

This shell is not uncommon in the drift, and owing to its dense texture is generally in good preservation. It ranges from the typical *Fusus tornatus* of Gould to *F. despectus* of Linnæus, as described by Fabricius, from Greenland, and shells of similar form from the British Crag are considered by S. Wood as varieties of *F. antiquus.*† Dr. P.

---

* Verrill calls this *Sipho lividus*, and holds that it is distinct from *Fusus Spitzbergensis*, Reeve. (J. F. W.)

† The *C. despectus* of Reeve, however, is a very different species, from the Arctic regions of the North Pacific.

P. Carpenter thinks that this and the British *F. antiquus* may prove to belong to one very variable species. The *F. despectus* is an Arctic form. The *F. tornatus* is recent, and is the form now found iu the Gulf and River St. Lawrence, where it is much larger and better developed than in the Pleistocene. *C. decemcostatus* is more southern.

Neptunea decemcostata. Say.

Fossil—Portland, Maine.
Recent—Magdalen Islands and Gaspé Bay (Whiteaves); coasts of Nova Scotia and New England.
This species has not yet been found in the Pleistocene of Canada, where it is represented by *C. torualus.* There are still two opinions as to whether Say's species is identical with *C. lyratus* (Mart.) = *Middendorffii* (Cooper) from the Pacific coast. The latter is variable, and graduates towards *tornatus* (Gould), but the living New England shells are tolerably constant in character.

Trophon scalariforme. Gould.

Fossil—Montreal ; Murray Bay ; Rivière-du-Loup ; Labrador.
Recent—Greenland (Hayes) ; Little Metis ; Murray Bay ; Nova Scotia (Willis) ; Gaspé and North Shore (Whiteaves).
It is a rare shell in the Pleistocene, but of large size and in good condition.

Trophon clathratum. Linn.

Fossil—Montreal ; Murray Bay ; Rivière-du-Loup ; Anticosti ; also glacial beds of Europe.
Recent—Rivière-du-Loup ; Little Metis ; Greenland and Arctic seas generally ; Labrador ; Gulf St. Lawrence (Whiteaves). The allied species or variety, *T. Gunneri*, has been found living at Gaspé by Whiteaves, and occurs at Little Metis, but not fossil as yet. British Columbia ; Alaska (Dall).

PROVINCE ANNULOSA.

Annulata.

Serpula vermicularis. Linn.

Fossil—Montreal ; Murray Bay ; Rivière.du-Loup.
A small species of Serpula, apparently the above, though perhaps the determination may be regarded as uncertain.

Vermilia serrula. Stimpson.

Fossil—Rivière-du-Loup, on shells.

Recent—Gulf St. Lawrence. It is quite likely the Greenland species identified by Fabricius with *Serpula triquetra.*

Spiochætopterus typus. Sars.

Fossil—Labrador (Packard).

Recent—Labrador (Packard) ; Norway (Sars).

Spirorbis glomerata (Stimpsoni, Verrill). Muller.

Fossil—Rivière-du-Loup ; Little Metis ; Labrador (Packard); Greenland (Fabr.) ; Gaspé. This species is *Spirorbis Stimpsoni* of Verrill, being regarded by him as distinct and new.

Spirorbis vitrea. Fabricius.

Fossil—Montreal ; Quebec ; Rivière-du-Loup ; Murray Bay. Very common on stones and shells.

Recent—Little Metis ; Greenland (Fabricius) ; Gulf St. Lawrence.

Spirorbis borealis. Daudin.

Fossil—Rivière-du-Loup, on shells.

Recent—Gulf St. Lawrence ; Greenland (Fabricius).

Spirorbis lucidus. Fleming.

Fossil—Rivière-du-Loup, on the inside of shells.

Recent—Gulf St. Lawrence ; Fishing Banks, American Coast (Gould).

Spirorbis carinata. Montague.

Fossil—Rivière-du-Loup, on shells.

This is a Spirorbis with one carina, found also in the Gulf of St. Lawrence, and possibly the same with the *S. contortuplicata* of Fabricius from Greenland ; Little Metis.

The beautiful *Spirorbis cancellata* of Fabricius, so common in the modern Gulf of St. Lawrence, and also in Greenland, has not yet been found in the Pleistocene.

Nereis pelagica. Linn.

Fossil—Green's Creek (collection of Mr. J. Stewart, Ottawa, and J.W. D.)

The specimens are in the nodules from this locality. They resemble at first sight whitish strips of calcareous matter about four inches in length and scarcely two lines in breadth. This strip of calcite is a

longitudinal section through the body of the worm, and shows nothing
of its external characters, and the somites of the body are indicated
only by the tufts of brown bristles or setæ at intervals along the sides.
In the best specimen these are in the middle portion of the body, from
a tenth to a twelfth of an inch apart. On the anterior segments they
are closer together, the body having apparently been contracted in that
part. Each foot, as indicated by the setæ—the soft parts having
entirely perished—seems to have had one strong spine and several
others, very fine and hair-like, in a separate bundle. When disengaged
from the matrix (which can easily be done by treating a small portion
with diluted acid) and examined microscopically, they seem to be
simple, nearly straight and pointed. Near what seems to be the anterior extremity are obscure indications of one of the horny mandibles.
These characters, as far as they go, would indicate a chætopod worm or
"sea centipede," and, of the species known to me on our coasts, they
resemble most closely that above named, which seems to be the *N. caeca*
of Fabricius, and is a common and widely distributed species in the
North Atlantic and Arctic seas.

PROVINCE ARTHROPODA.

Class IV.—Crustacea.

The most abundant species are bivalve Entomostraca,
which occur in great numbers in the Leda clay, associated
with Foraminifera. The species in my collection have
been kindly determined by Mr. J. S. Brady, who enumerates
the following:—

Cythere MacChesneyi, nov. sp.
" *Dawsoni* (Brady).
— " *globulifera* (Brady).
" *Logani*, nov. sp.
Cytheridea papillosa (Bosquet).
" *punctillata* (Brady).
" *Sorbyana* (Jones).
" *Mulleri*.
Cytherura Robertsoni (Brady).
Cytheropteron complanatum, nov. sp.
" *inflatum* (B., C., and R., MS.)
" *angulatum* (B., C., and R., MS.)
Eucythere argus.

As the paper was re-printed in the *Canadian Naturalist* (Vol. V., N.S.) it is unnecessary to notice these species further here, except to state that out of twenty-nine species of recent Ostracods obtained by Mr. Brady from material from the Gulf St. Lawrence, furnished by me, thirteen have been recognized in the Pleistocene of Canada and Maine. It is further remarkable that out

FOSSILS.—PLATE VII.

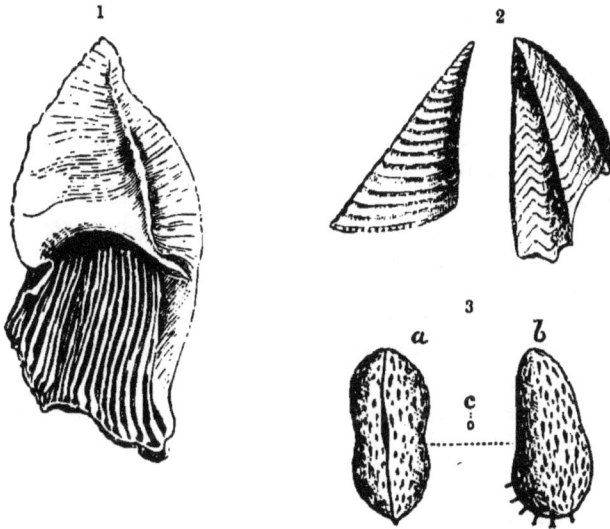

Pleistocene Crustacea.—1, *Balanus Hameri* (one of the body valves); 2, *B. Hameri* (opercular valves); 3, *Cytheridea Mulleri*—(*a*) valves united; (*b*) left valve; (*c*) natural size.

of thirty-three fossil species from Maine and Canada, no less than twenty-three occur in the Scottish glacial beds and twenty-five are living in the British seas, while six are new species.

Balanus Hameri. Ascanius.

Fossil—Montreal; St. Nicholas; Quebec; Anticosti; Rivière-du-Loup; also, Uddevalla; Russia (Murchison); Greenland (Spengler).

Recent—Coast of Nova Scotia. I have obtained specimens from Mr. Downes of Halifax, but have not elsewhere seen the species recent. It is *B. Uddevallensis* of lists of Scandinavian fossils and *B. tulipa* of Muller. It is a widely diffused Arctic and North Atlantic species.

This Acorn-shell is very abundant at Rivière-du-Loup, and fine specimens are found entire, attached to stones and boulders in the boulder-clay.

Very fine specimens are also obtained at River Beaudette, about 34 miles west of Montreal. This locality is noteworthy as being further west than the others mentioned. The specimens are also interesting from their remarkable perfection and the large masses which they form, some of which contain as many as a dozen individuals attached to each other. They were collected by Mr. A. W. McNown, of Rivière Beaudette, and by Mr. Stanton, C.E., of Lancaster.

The animals seem to have been covered, when living, with an irruption of sand, for the opercular valves of many of them are still in place, and, owing to a slight infiltration of calcareous matter, the radial plates and opercular valves have been cemented together, which accounts for their perfect preservation. It is to be observed, however, that the shells of Balani are composed of a remarkably dense and indestructible calcium carbonate, much less perishable than the shells of most mollusks.

The original attachments of the animals, so far as observed, have been on pebbles on the surface of clay, and as these afforded space only for one or two individuals, the young were obliged to attach themselves to the old in successive generations, forming grotesque groups, which still remain entire. They are associated with *Saxicava rugosa, Mya arenaria* and *Macoma Grœnlandica.*

Balanus porcatus. DaCosta.

Fossil—Beauport; glacial beds of Europe.

Recent—Gulf St. Lawrence, and coast of New England; Labrador (Packard); and Arctic and Northern seas generally. It is no doubt *Lepas balanus* of Fabricius from Greenland.

Much more rare in the Pleistocene than the preceding species.

Balanus crenatus. Brug.

Fossil—Montreal; Quebec; Rivière-du-Loup; Anticosti; St. John, N.B. (Matthew); Labrador (Packard); Portland, Maine; glacial beds of Europe; Vancouver Island (G. M. D.)

Recent—Arctic and northern seas, Greenland; Gulf St. Lawrence and American coast. It seems to be *Lepas balanaris* of Fabricius, from Greenland.

Balanus balanoides.

Fossil—Portland, Maine.   Not yet in Canada.

Eupagurus Bernhardus? Fabricius.

Fossil—Rivière-du-Loup.   A small specimen in a *Turritella* may be the young of this common species.

Hyas coarctata. Leach.

Fossil—Rivière-du-Loup.   A few claws only found, but evidently of this common Gulf of St. Lawrence species.

Estheria Dawsoni. Packard.

Fossil—Green's Creek, Ottawa.

A new species found in the nodules containing fishes, &c., and described by Packard as follows* :—

This Estheria is entirely unlike any northern American or European species, differing decidedly from *Estheria morsei* or *E. caldwelli* and *E, clarkii*. It rather approaches *E. jonesii* from Cuba in the form of the shell and style of marking of the valves. It does not resemble closely any of the fossil forms figured in Jones' Monograph of Fossil Estheriæ. The markings, however, present some resemblances to *E. middendorfii* Jones, but differs in the want of anastomosing cross wrinkles between the ridges.

One valve and portions of others were preserved ; but none of them show the beaks (umbones), though the form of the remainder of the shell indicates that they were situated nearer the middle of the valve than usual—*i.e.*, between the middle and the anterior third of the shell. The shell is deep, probably more so than in *E. jonesii*, though the valves have evidently been flattened and somewhat distorted by pressure, but apparently the head-end was more truncated than in *E. jonesii*, as the edge of the shell and the parallel lines (or ridges) of growth along the head-end are, below, bent at right angles to the lower edge of the shell. The raised lines of growth are very numerous and near together ; they are of nearly the same distance apart above, near the beaks, as on the lower edge. The very numerous lines of growth are thrown up into high sharp ridges, the edges of which are often rough, finely granulated, and often the valleys between are rugose on the surface. In one or two places a row of papillæ for the insertion of spinules may be seen where the shell has been well preserved, and between many of the lines of growth there are irregular superficial ridges.   Length, 10 mm. ; depth, 7.5 mm.

---

* *American Naturalist*, June, 1881.

The valve is evidently that of an Estheria, much truncated anteriorly, and with the lines of growth much thicker, higher and closer together than in any North American species known to us, and may prove, when better specimens are found, to be allied to the tertiary Siberian *E. middendorfi.*

Insecta.

Fornax ledensis. Scudder.

This is an elytron of a beetle in a nodule from Green's Creek, along with a skeleton of *Mallotus Villosus.* Scudder, who has kindly examined it, regards it as representing a new species allied to *F. calceatus* of North America. It has been described by Dr. S. in his volume on Tertiary insects.

Scudder in his work on Fossil Insects and previously in Reports to the U. S. Geological Survey, notices the insects collected by Hinde in the interglacial beds at Scarborough, on Lake Ontario. He regards these insects as extinct species, but nearly related to modern temperate forms, and in no respect an Arctic assemblage. This agrees with the evidence of the fossil plants.

PROVINCE VERTEBRATA.

The vertebrate animals of the Pleistocene are few; and we can scarcely include in this formation the Mastodon or Mammoth, and their contemporaries, as their remains, so far as known in Canada, are rather Post-glacial or Modern. The fishes are mostly from nodules in the Leda clay, found at and near Green's Creek on the Ottawa River, and are ordinary northern species.

Class Pisces.

Mallotus villosus. Cuvier.

The common capelin is found in nodules at Green's Creek on the Ottawa. (Plate VII.)

Osmerus mordax. Gill.

An imperfect skeleton, apparently referable to the smelt, Green's Creek, Ottawa.

Cottus (Centrodermichthys) uncinatus. Reinhardt.

Fossil—Nodules from Green's Creek, collection of Mr. J. Stewart, Ottawa, and of J. W. D.

FOSSILS.—PLATE VIII.

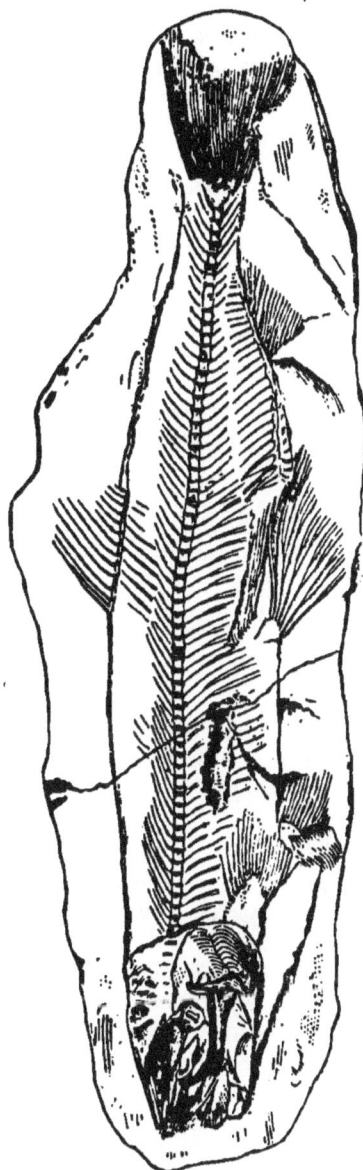

1. Mallotus villosus (Capelin), in nodule.

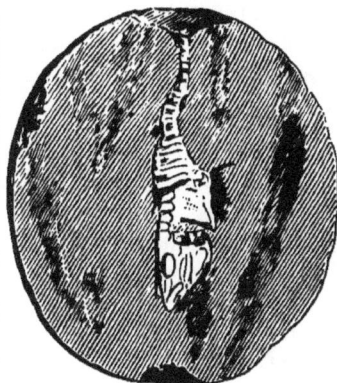

2, Gasterosteus, in nodule.

There have been in my collections for some time two specimens of these nodules, which appear to contain the skeletons of some species of *Cottus* or Sculpin. They are, however, imperfectly preserved, so that I have been unable to identify the species. Recently, Mr. J. Stewart, of Ottawa, has kindly placed in my hands a better preserved specimen, showing more especially the pre-opercular spines and pectoral fins in comparatively good preservation, and with the help of these I think I can identify the species, notwithstanding the confusion which at present seems to reign as to our North American cottoids.

The characters of the hooked spines and of the pectoral fin seem to identify this specimen with *Cottus* (*Centrodermichthys*) *uncinatus* of Gunther's British Museum catalogue. This is *C. uncinatus* of Reinhardt, and *Icelus uncinatus* of Kroyer and Gill. I feel convinced, also, that it must be the *Cottus gobio* of Fabricius, though this is usually identified with *C. (Gymnacanthus) tricuspis* of Reinhardt, a very distinct species. *Cottus uncinatus* occurs in Greenland and in deeper water as far south as New England, according to Jordan, who creates for it a new genus (*Artediellus*).*

The total length of the specimen without the caudal fin, which is absent, is four inches, of which the head measures one inch. It belongs to the collection of Mr. Stewart. The other and less perfect specimens, which I refer to the same species, are in the Peter Redpath Museum.

Cyclopterus lumpus. Linn.

The lump sucker occurs in nodules at the same place.

Gasterosteus aculeatus? L.

In nodules at the same place, found by Sheriff Dickson. It closely resembles the two-spined stickleback of the Gulf St. Lawrence, but is not sufficiently perfect for detailed description.

Salmo salar? Linn.

Fossil—A head apparently referable to this species in a nodule from Goose River, north shore of River St. Lawrence.

Vertebræ and other fragments of fishes not determinable have been found at Rivière-du-Loup and other places.

Class Aves.

A few specimens of feathers have been preserved in nodules at Green's Creek. They have apparently belonged to small wading birds.

Class Mammalia.

Phoca (Pagophilus) Grœnlandica. Muller.

A nearly complete skeleton of this species, found some years ago in the Leda clay near Montreal, is now in the collection of the Geological Survey of Canada at Ottawa. Detached bones, also found near Montreal, are in the Peter Redpath Museum of McGill University.

More than twenty years ago, Mr. Billings, then at Ottawa, obtained a nodule with certain bones enclosed in it from the Pleistocene clays of Green's Creek, on the Ottawa, which have afforded so many beautiful specimens of the Capelin and other fishes, and also of marine shells of northern and cold water types. Mr. Billings regarded the bones as those of the limbs of "a small animal of aquatic habits," but, not being

---

* Catalogue of Fishes, Fish Commission Reports.

able to determine the species, sent the specimen to Dr. Leidy, of Phila-delphia. He recognized the bones as those of the hinder extremity of a young seal, but of what species was uncertain. A good figure and description were published in the first volume of the *Naturalist* in 1856. No further information bearing directly on this fossil was secured until the discovery some years ago of a jaw bone of a young individual of this species by Sir J. Grant. It is the left ramus of the lower jaw of a young seal, containing a canine and four molar teeth, with an impres-sion of the fifth. It enables us to affirm that the species is *Phoca Grœnlandica* (*Pagophilus Grœnlandicus* of Gray's catalogue)—the common Greenland seal, and it is of such size that it may have belonged to the same individual which furnished the bones described in 1856, or at least to an animal of the same species and of similar age.

Beluga catodon (*Delphinapterus leucas*, Pallas ; *Beluga vermontana*, Thompson).

Bones of this species have been found at Rivière-du-Loup and at Montreal, in the Saxicava sand near Cornwall (Billings) and in the same deposit near Bathurst (Gilpin and Honeyman). There seems no good reason to believe that the *B. Vermontana* of Thompson from the Pleistocene of Vermont is distinct from this species.

Megaptera longimana. Gray.

Portions of a skeleton of this species were found in 1882 in a ballast pit on the Canadian Pacific Railway, three miles north of Smith's Falls, in Ontario, 31 miles north of the St. Lawrence River. They were im-bedded in gravel along with shells of *Tellina Grœnlandica*, apparently on a beach of the Pleistocene period at an elevation of 440 feet above the sea, which corresponds nearly with one of the principal sea coast terraces on the Montreal mountain and other parts of the St. Lawrence valley.

The specimens obtained were presented by Mr. A. Baker, of the C. P. Railway, to the Peter Redpath Museum. They consist of a lumbar and dorsal vertebra and a rib, and correspond with these bones in the species above named, which seems to be *Balœna boops* of Fabricius. It is still found in the Gulf of St. Lawrence, and is more disposed than the other large whales to extend its excursions some distance into the estuary of the St. Lawrence and other narrow seas.

Belcher (Last of the Arctic Voyages, 1853) mentions the occurrence of the bones of a large whale imbedded in clay at Mount Parker, at an elevation of 500 feet ; and at Cape D'Israeli a similar specimen at about the same elevation. On the Lower St. Lawrence, bones of large

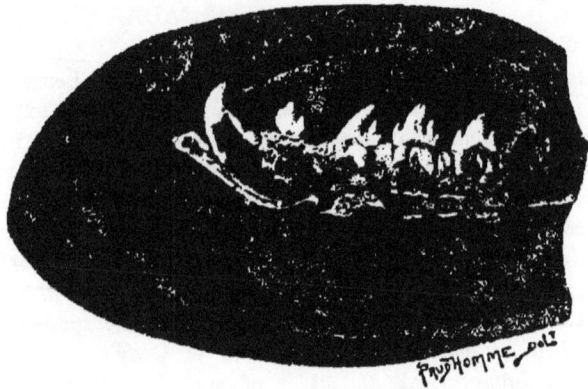

Pagophilus Grœnlandicus, Muller.

The above illustration, for which I am indebted to Mr. H. M. Ami, F.G.S., represents the jaw of this species (referred to at p. 268), in the collection of Sir James Grant, of Ottawa.

Mr. Ami has also kindly sent me the names of the following species occurring at Green's Creek, and which are not mentioned in the above lists as from that locality: *Asterias,* sp., *Balanus crenatus, Mytilus edulis, Macoma fragilis, Natica affinis,* two additional insects, *Tenebrio calculensis,* and *Byrrhus Ottawaensis.*

whales are not infrequent on the lower marine terraces, and are reported as occurring also on the higher terraces, but this I have not verified by personal observation. They probably belong either to the "Humpback" or to the "Finner" whale, both of which are occasionally present in the Lower St. Lawrence, and are said in former times to have been more numerous. I secured last summer (1891) a large jaw-bone found in digging a cellar in the shelly gravel of the lower terrace at Metis. It is now in the Peter Redpath Museum.

THE ARCTIC BASIN.

It may be of interest to add here a list of the species recognized by Jeffreys in the collections of Capt. Fielden in the Pleistocene of Grinnell Land and North Greenland (*Ann. and Mag. Nat. Hist.*, 1877 p. 230 ; *Zoologist*, 1877, pp. 485–440). It would appear that these shells are found at various elevations, from near the sea level to about 1,200 feet.

Conchifera.

Pecten Grœnlandicus. Sowerby.

Pecten Islandicus. L.

Leda pernula. Muller.

Leda arctica. Gray.

Leda frigida (Torrell) = *Yoldia Nana* (Sars.)

Axinus flexuosus. Montague. Var. *Gouldii.*

Arca glacialis. Gray.

Cardium Islandicum. Chemnitz.

Chemnitz, Conch. Cab., Vol. VI., p. 200, tab. 19, figs. 195, 196. Circumpolar; frequent in Post-tertiary deposits throughout the north of Europe and America.

Astarte borealis. Chemnitz.

Astarte fabula. Reeve.

This species is probably the *Nicania Banksii* of Leach, MS., which was figured by the late Mr. G. B. Sowerby in his Supplement to Gray's "Mollusca of Beechey's Voyage" (1839), pl. XLIV., fig. 10, as "*Astarte Banksii?* (Gray) in Brit. Mus." Möller included it in his list of

Greenland Mollusca under the name "*Nicania Banksii*, Sab." Reeve's publication was in 1855. Sowerby's figure, although it represents the shape, does not show the peculiar sculpture of *A. fabula*. See also *A. Crebricostata* (Forbes) and *A. Richardsonii*. Reeve.

Tellina calcarea. Chemn.

Tellina calcarea. Chemn., VI., p. 140, tab. 13, fig. 136. For the synonymy and range of this common Arctic shell and "glacial" fossil, see "British Conchology," Vol. II., pp. 389, 390, and Vol. V., p. 187.

Thracia obliqua. Jeffr., sp. n.

A valve measuring an inch and six-tenths in breadth by an inch and one-tenth in length. It is distinguishable from *Thracia* (*Amphidesma*) *truncata* of Brown, = *T. myopsis* (Beck), in having a more oblique or twisted shape, a straight instead of rounded margin in front, and a more gradual or less abrupt slope to each side ; the truncature at the posterior side is broad and regularly curved ; and the surface is puckered, as in *Mya truncata*. It wants the flexuosity of *T. pubescens*, but resembles in its outline that species more than *T. truncata*. *T. septentrionalis*, Jeffr. (*truncata* of Mighels and Adams), differs in shape and texture from all the above-named species. Having, however, seen but a single valve, I will not insist on this constituting a new species.

Mya truncata, L., (var. *Uddevallensis*).

Saxicava rugosa. L.

Neaera subtorta. Sars.

Solenoconchia.

Siphodentalium vitreum.

Gastropoda.

Trochus (*Margarita*) *umbilicalis*. Brod. and Low.

Trichotropis borealis. Brod. and Low.

Buccinum tenue.

Buccinum hydrophanum. Hancock.

Trophon clathratus. L.

Pleurotoma tenuicostata. Say.

P. Exarata. Möller.

P. Trevelyana. Fischer.

Cylichna alba. Brown.

Actinozoa.

Funiculina quadrangularis.

Mr. Norman has examined these organisms, and favoured me with the following memorandum :—

"*Funiculina quadrangularis* (Pallas) = *Pavonaria quadrangularis*. Johnston.

"Fragments of the full-grown quadrangular calcareous skeleton-rods. They are in good condition, and much more recent-looking and less decayed than a similar rod which I dredged two months ago to the N.E. of the 'Maiden Rock' near Oban. There can be no doubt that this Oban specimen was 'recent'; for although I did not dredge it living, it was close to this locality that Mr. McAndrew obtained the first known British example of this species. *Funiculina quadrangularis* is at present known to range from the Adriatic Sea (Kölliker) to the Minch ('Porcupine' Expedition, 1869) on our own coast, and Kattegat in the Scandinavian Seas (Malm)."

Foraminifera.

Cornuspira foliacea. Philippi.

Marine Algæ.

Melobesia polymorpha.

FOSSIL PLANTS.

The first locality where fossil plants in any considerable number were obtained, was Green's creek on the Ottawa, where they owe their preservation to the nodules of calcareous matter that have enclosed delicate specimens, which otherwise could not have been secured from the soft Leda clay in which the nodules are enclosed. They are associated with *Leda arctica* and with skeletons of *Mallotus* and other fishes. In addition to specimens collected by myself, I have examined the collections made by the late Rev. Mr. Bell of L'Orignal, those of the late Sheriff Dickson, and those of the Geological Survey. The whole were described in my paper in the *Canadian Naturalist* for February, 1866, and included nine or ten species of phaenogams, a moss, and an alga. Subsequently, additional specimens from this place were collected by the late Mr. J. S. Miller, and by Mr. J. Stewart of

FOSSILS.—PLATE IX.

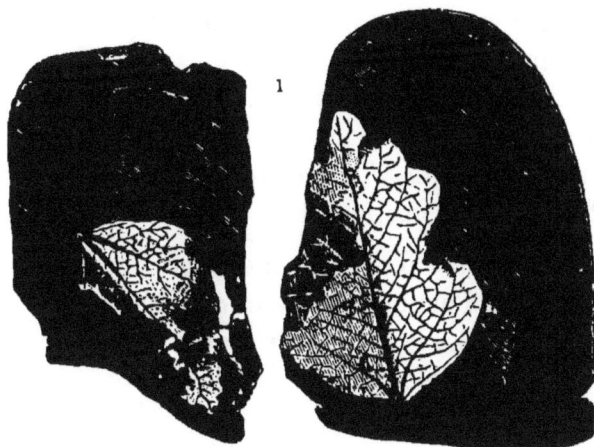

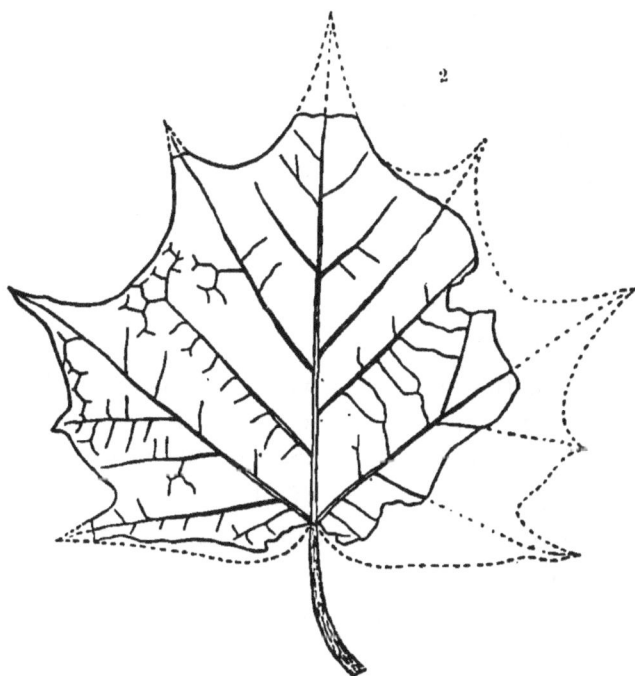

1, *Populus balsamifera.*   2, *Acer Pleistocenicum.*

Ottawa, and were placed in my hands; while specimens of wood found at different times in the Leda clay of Montreal were also placed by me in the Peter Redpath museum.

The interesting deposits at Scarboro' heights and elsewhere on Lake Ontario were described by Dr. J. G. Hinde in the *Canadian Journal* in 1877, and he notices the following plants as found by him:

> Wood of pine and cedar.
> Portions of leaves of rushes, etc.
> Seeds of various plants.
> *Hypnum commutatum.*
> *H. revolvens.*
> *Fontinalis.*
> *Bryum.*
> *Chara*, sp.

More recently Mr. J. Townsend, of Toronto, was so fortunate as to find leaves and fragments of wood, with shells of *Melania* and *Cyclas*, in beds apparently of the same age, in excavations in progress on the River Don, at Toronto.

The section observed at this place is given as follows by Mr. Townsend:

The locality of the principal vegetable specimens was 150 feet from the bank of the Don, and in a cutting 70 feet deep. The section showed 26 feet of fine light-colored sand, with layers of clay at bottom. Below this were 24 feet of tough stratified blue clay, the "Erie clay" of the region. At the base of this clay is a seam of reddish ferruginous sand, about three feet thick, and with argillaceous nodules, in which was the maple leaf de-

19

scribed by Professor Penhallow. Below this sand were
16 feet of alternating sand and dark-colored clay, with
fresh-water shells and wood. Below this was the blue
till resting on the surface of the Hudson river beds. In
this section the upper boulder-clay of Hinde's section is
not represented, but only the lower groups as given in his
table. The upper boulder-clay is, however, seen on higher
ground in the vicinity.

Dr. J. W. Spencer, who has studied this locality, as
well as the whole north shore of Lake Ontario, writes to
me that he regards the earthy sand holding wood and
fresh-water shells as equivalent to Hinde's "interglacial"
beds at Scarboro' heights, and the overlying clay as the
so-called "Erie clay," over which, as above stated, is the
upper boulder deposit which, in the vicinity of Toronto,
has many Laurentian boulders.

Observations have been made on the interglacial beds
of the West by Dr. G. M. Dawson, and are recorded
in his reports on the 49th Parallel, and on the geology of
the Bow and Belly rivers, and in a paper on borings made
in Manitoba and the North-west Territories, in Vol. IV.
of the Transactions of the Royal Society of Canada; and
he has placed in my hands specimens of peat and wood
from those regions. In one locality on the Belly river he
finds a bed of interglacial peat, hardened by pressure in
such a manner as to assume the appearance of a lignite.

In addition to the vegetable remains found as above
stated in the "forest beds" or "interglacial" deposits,
trunks of trees and vegetable fragments occur in the
boulder-clays themselves, indicating either the partial
destruction of the older interglacial bed and the mixture
of its débris with glacial deposits, or the enclosure of
drift-wood in the latter in the manner now so common

in the arctic regions, and described by so many arctic explorers.*

One of the most marked illustrations is that of the boring at Solsgirth, in Manitoba, on the Manitoba and North-western railway, and at an elevation of 1,757 feet above the sea.† At this place the section is as follows:

		Feet.
1.	Loam	2
2.	Hard blue clay and gravel	42
3.	Hard blue clay and stones	10
4.	Hard yellow "hard pan"	12
5.	Softer bluish clay	16
6.	" " "	74
7.	Sand with water	..
8.	Blue clay with stones	136
9.	Gray clay or shale (Cretaceous?)	68
		360

Fragments of wood, more or less decayed and compressed, were obtained from depths of 96, 107, 120 and 135 feet from the surface. They were thus distributed through a considerable thickness of the clay rather than in a distinct interglacial deposit. It is to be observed, however, they were included within the central part characterized as a softer blue clay, between two beds apparently harder and more stony.

Additional specimens from this place have recently been obtained by Mr. J. B. Tyrrell, of the Geological Survey of Canada, and have been kindly communicated to me. Mr. Tyrrell has also found vegetable remains in a

---

* See Manual of the Natural History, Geology and Physics of Greenland, by Professor T. R. Jones, issued by the Royal Society of London, 1875, index—"Driftwood."

† Dr. G. M. Dawson, Trans. Royal Society Canada, Vol. IV., 1887, sec. IV., p. 91. et seq.

bed under the boulder-clay at Rolling river, Manitoba, which are noticed in Professor Penhallow's paper. They were accompanied with fresh-water shells of the following species, determined by Mr. Whiteaves, F.G.S., Palæontologist to the Geological Survey of Canada :

>*Lymnea catascopium?*, variety with very short spire.
>*Valvata tricarinata*, and a keelless variety.
>*Amnicola porata ?*
>*Planorbis parvus ?*
>*P. bicarinatus.*
>*Pisidium abditum.*
>*Sphærium striatinum.*

With these was the centrum of a vertebra of a small fish.

Dr. G. M. Dawson has also found fragments of wood at Skidegate, Queen Charlotte Islands, in boulder-clay, associated with shells of *Leda*, etc.

As elsewhere stated, at River Inhabitants, in Cape Breton, there is an indurated peat with branches of *Taxus* and remains of swamp plants *below* the boulder-clay.

The whole of the above collections have been placed in the hands of Prof. Penhallow, of McGill University, for revision and determination, and his results have been published in the Bulletin of the Geological Society of America, Vol. I., to which reference may be made for details.

The whole number of Canadian species has thus been raised to 33, as follows :—

1. *Asimina triloba,* Dunal. Don River, Toronto (Townsend).
2. *Brasenia peltata,* Pursh. Green's Creek nodules (Miller).
3. *Drosera rotundifolia,* L. Green's Creek, Ottawa (J. W. Dawson).*

---

* Collection of Sir William Dawson in Peter Redpath Museum.

4. *Acer saccharinum*, Wang. Green's Creek, Ottawa (J.W. Dawson).
5. *Acer pleistocenicum*, sp. nov. Don River, Toronto (Townsend).
6. *Potentilla anserina*, L.
    Green's Creek, Ottawa (J. W. Dawson and Miller).
7. *Gaylussacia resinosa*, Torr. and Gray.
    Green's Creek, Ottawa (J. W. Dawson).
8. *Menyanthes trifoliata*, L. Leda clays, Montreal.*
9. *Ulmus racemosa*, Thomas, Don River, Toronto (Townsend).
10. *Populus balsamifera*, L. Green's Creek, Ottawa (J. W. Dawson).
11. *Populus grandidentata*, Michx.
    Leda clays, Montreal (Weston).
    Green's Creek nodules (Stewart).
12. *Picea alba*, Link. Bloomington, Ill. (Andrews).
13. *Larix americana*, Michx. *Leda* clays, Montreal (Weston).
14. *Thuya occidentalis*, L.
    Leda clays, Montreal (J. W. Dawson).
    Leda River, Manitoba (Dr. G. M. Dawson).
    Marietta, Ohio (Newberry).
15. *Taxus baccata*, L.
    Don River, Toronto (Townsend).
    Solsgirth, Manitoba (G. M. Dawson and Tyrrell).
    Rolling River, Manitoba (Tyrrell).
    Cape Breton (Sir William Dawson).
    Bloomington, Ill. (Andrews).
16. *Potamogeton perfoliatus*, L. Green's Creek, Ottawa (J. W. Dawson).
17. *Potamogeton pusillus*, L. Green's Creek, Ottawa (J. W. Dawson).
18. *Potamogeton rutilans*(?),Wolfgang. Green's Creek nodule(Stewart).
19. *Elodea canadensis* (?), Michx. Rolling River, Manitoba (Tyrrell).
20. *Vallisneria* (?). Rolling River, Manitoba (Tyrrell).
21. *Carex magellanica*, Lamarck.
    Green's Creek nodules, Ottawa (Miller and Stewart).
22. *Oryzopsis asperifolia*,Michx. Green's Creek,Ottawa(J.W. Dawson).
23. *Bromus ciliatus* (?), L. Green's Creek, Ottawa (Miller).
24. *Equisetum sylvaticum* (?), L. Green's Creek nodules (Stewart).
25. *Equisetum limosum* (?), L. Green's Creek, nodules (Stewart).
26. *Equisetum scirpoides*,Michx. Green's Creek,Ottawa(J.W.Dawson).
27. *Fontinalis* (?), sp. Green's Creek, Ottawa (J. W. Dawson).
28. *Fucus*, sp. Green's Creek, Ottawa (J. W. Dawson).
29. *Navicula lata*. Rolling River, Manitoba.
30. *Encyonema prostratum*. Rolling River, Manitoba.

* Collection of Sir William Dawson in Peter Redpath Museum.

31. *Denticula lauta.* Rolling River, Manitoba.
32. *Licmophora* (?). Rolling River, Manitoba.
33. *Cocconeis.* Rolling River, Manitoba.

None of the plants above mentioned are properly arctic in their distribution, and the assemblage may be characterized as a selection from the present Canadian flora of some of the more hardy species having the most northern range. Green's creek is in the central part of Canada, near to the parallel of 46°, and an accidental selection from its present flora, though it might contain the same species found in the nodules, would certainly include with these, or instead of some of them, more southern forms. More especially the balsam poplar, though that tree occurs plentifully on the Ottawa, would not be so predominant. But such an assemblage of drift plants might be furnished by any American stream flowing in the latitude of 50° to 55° north. If a stream flowing to the north it might deposit these plants in still more northern latitudes, as the McKenzie river does now. If flowing to the south, it might deposit them to the south of 50°, In the case of the Ottawa, the plants could not have been derived from a more southern locality, nor probably from one very far to the north. We may therefore safely assume that the refrigeration indicated by these plants would place the region bordering the Ottawa in nearly the same position with that of the south coast of Labrador fronting on the Gulf of St. Lawrence, at present. The absence of all the more arctic species occurring in Labrador, should perhaps induce us to infer a somewhat more mild climate than this.

The climatic indications afforded by these plants are not dissimilar from those furnished by a consideration of the marine fauna of the period of the Leda clay.

SUMMARY OF FOSSILS.

The above lists include, in all, about 240 species, distributed as follows :*

Plants...................................	33
Animals —Protozoa, etc................	21
Echinodermata.....................	7
Mollusca.........................	142
Annulosa and Arthropoda...........	30
Vertebrata.......................	7
	240

The whole of the marine species, with two or three exceptions, may be affirmed to be living northern or Arctic forms, belonging, in the case of the marine species, to moderate depths, or varying from the littoral zone to say 100 fathoms. The assemblage is identical with that of the northern part of the gulf of St. Lawrence and Labrador coast at present, and differs merely in the presence in the modern gulf of a few more southern forms, especially in its southern part, where the fauna is of a New England type, whereas that of the Pleistocene may be characterized as Labradorian, or at least as corresponding to that part of the gulf of St. Lawrence now invaded by the Labrador cold current.

I would call attention in this connection to the number of species recorded as recent on the evidence of my own dredgings in the lower St. Lawrence at Metis, Rivière-du-Loup, Murray bay, and Kamouraska. In point of fact nearly all the marine species of the Leda clay and Saxicava sand are still living on the coasts opposite the points where the fossils occur. It is to be observed, however, that in the modern river and gulf they are associated with

* Exclusive of a few fresh-water species mentioned in the text, and of which I have not seen specimens.

some living species of less boreal forms, not found in the Pleistocene beds.

Some of the species, it will be seen, are of very wide distribution in the modern seas, occurring in the Pacific as well as in the Atlantic.

As might have been anticipated from the relations of the modern marine fauna, the species of the Canadian Pleistocene are in great part identical with those of the Greenland seas and of Scandinavia, where, however, there are many species not found in our Pleistocene. The Pleistocene fauna of Canada is still more closely allied to that of the deposits of similar age in Britain and in Norway. Change of climate, as I have shown in previous pages, has been much more extensive on the east than on the west side of the Atlantic, owing to the distribution of warm and cold currents, resulting from the differing elevation of the land.

It cannot be assumed that the fauna of the older part of the Canadian Pleistocene is different to any great extent from that of the more modern part. Such difference as exists seems to depend on variations of depth or on a gradual amelioration of climate. The shells of the lower boulder-clay, and of those more inland and elevated portions of the beds which may be regarded as older than those of the lower terraces near the coast, are undoubtedly more arctic in character. In some localities they are confined to a few species such as occur in the permanently ice-laden seas of Spitzbergen. The amelioration of the climate seems to have kept pace with the gradual elevation of the land, which threw the cold ice-bearing arctic currents from its surface, and exposed a larger area to the direct action of solar heat, and also probably determined the flow of the waters of the Gulf

Stream into the North Atlantic. By these causes the summer heat was increased, the winds both from the land and sea were raised in temperature, and the heavy northern ice was led out into the Atlantic, to be melted by the Gulf Stream, instead of being drifted to the south-west over the lower levels of the continent. Still the cold arctic currents entering by the straits of Belle-isle and the accumulation of ice and snow in winter, are sufficient to enable the old arctic fauna to maintain itself on the northern side of the gulf of St. Lawrence, and to extend as far as the latitudes of Murray bay and Gaspé. South of Gaspé we have the warmer New England fauna of Northumberland strait. I may add that some of the varietal peculiarities of the Pleistocene fauna in comparison with that of the St. Lawrence river, indicate a considerable influx of fresh water, derived possibly from melting ice and snow.

MAN IN CANADA.

No remains of man or of his works have yet been found in the Pleistocene of Canada, though discoveries of implements have been recorded from alluvial deposits at depths which indicate a considerable historical antiquity ; still they do not go farther back than the Modern period, properly so called. Nor am I aware that human remains have been found in those early Modern gravels, alluvia, and sub-soils of bogs, which seem to be the repositories of the remains of the Mastodon and Mammoth.

The Post-glacial, or early Modern period in Canada, was, as indicated in a previous chapter, characterized by an elevation of the land to a greater height than at present, accompanied with a marked amelioration of climate, connected, perhaps, with the narrowing of those northern channels which supply drift ice to the north Atlantic, and with a wider heating-surface of low land. In this respect eastern America corresponded with Europe, and a similar mammalian fauna overspread both sides of the Atlantic. In this "Second Continental" period, as it has been called, man certainly appeared in Europe, and not improbably in America, though this may as yet be regarded as uncertain.

Every reader of the scientific journals of the United States must be aware of the numerous finds of "palæo-lithic" implements in "glacial" gravels. I have endeavoured to show, in a work published several years ago,* how much doubt attaches to the reports of these discoveries, and how much such of the "palæoliths" as appear to be the work of man resemble the rougher tools and rejectamenta of the modern Indians. But since the publication of that work, so great a number of "finds" have been recorded, that, despite their individual impro-bability, one was almost overwhelmed by the coincidence of so many witnesses. Now, however, a new aspect has been given to the question by Mr. W. H. Holmes, of the American Geological Survey, who has published his observations in the *American Journal of Anthropology* and elsewhere.†

One of the most widely known examples was that of Trenton on the Delaware, where there was a bed of gravel alleged to be Pleistocene, and which seemed to contain enough of "palæolithic," implements to stock all the

---

* "Fossil Men," Hodder & Stoughton, London, 1880.
† *Science*, Nov., 1892. *Journal of Geology*, 1893.

museums in the world. The evidence of age was not, however, satisfactory in a geological point of view, and Holmes, with the aid of a deep excavation made for a city sewer, has shown that the supposed implements do not belong to the undisturbed gravel, but merely to a talus of loose debris lying against it, and to which modern Indians resorted to find material for implements, and left behind them rejected or unfinished pieces. This alleged discovery has therefore no geological or anthropological significance. The same acute and industrious observer has inquired into a number of similar cases in different parts of the United States, and finds all liable to objections on the above grounds, except in a few cases when the alleged implements are probably not artificial. These observations not only dispose, for the present at least, of palæolithic man in America, but they suggest the propriety of a revision of the whole doctrine of " palæolithic " and "neolithic" implements as held in Great Britain and elsewhere. Such distinctions are often founded on forms which may quite as well represent merely local or temporary exigencies, or the debris of old workshops, as any difference of time or culture. All this I reasoned out many years ago on the basis of American analogies, but the Lyellian doctrine of modern causes as explaining ancient facts seems as yet to have too little place in the science of Anthropology. It may be added that Wright, in recent papers, attempts to defend some of the " palæo- lithic " finds against Holmes's criticisms; and a somewhat active controversy is still in progress. The evidence, however, for the Pleistocene age of any of the genuine implements seems too uncertain to be accepted at present. All that can be affirmed is that there is a certain probability that men of the American type existed in America in the Post-glacial or early Anthropic period, and may have been contemporary with the Mastodon and the gigantic animals now extinct. This subject, however, is not within the scope of the present work; and I have discussed it sufficiently elsewhere.*

* "Fossil Men." "Modern Science in Bible Lands."

Map of the Pleistocene Cordilleran Glacier, after Dr. G. M. Dawson.—The short
curved lines indicate the glacial margin and movement. The long black
line on East side of the Rocky Mountains, the limit of boulders from the
Laurentian.

CHAPTER VII.

GENERAL CONCLUSIONS.

These have, perhaps, been sufficiently indicated in an incidental manner in the preceding pages; but it may be well here to note some results of a less special character and bearing on larger biological and cosmical questions.

With reference to the life of the Pleistocene period, one can scarcely fail to observe that, whatever may have been the lapse of geological time from the period of the oldest boulder-clay to that in which we live, and great though the climatal and geographical changes have been, we cannot affirm that any change, even of varietal value, has taken place in any of the species of the above lists. This appears to me a fact of extreme significance with reference to theories of the modification of species in geological time. No geologist doubts that the Pleistocene was a period of considerable duration. The great elevations and depressions of the land, the extensive erosions, the wide and thick beds of sediment, all testify to the lapse of time. The changes which occurred were fruitful in modifications of depth and temperature. Deep waters were shallowed, and the sea overflowed areas of land. The temperature of the waters changed greatly, so that the geographical distribution of marine animals was materially affected, and they have had to make important changes of habitat, while some of them have so extended their range as to be found on both sides of the North Pacific and North Atlantic. Yet all the Pleistocene species survive, and this without change. Even variable forms like the species of *Buccinum* and *Astarte* show the same range of variation in the Pleistocene as in the modern, and though some varieties have changed their geographical position, they have not changed their character. These changes of geographical position are also very significant, as they seem to show that arctic and temperate varieties are readily convertible into each other when the temperature of the water changes, but revert to the old forms on restoration of the old conditions.* This result is obviously independent of imperfection of the geological record, because there is no reason to doubt that these species have continuously occupied the North Atlantic area, and we have great abundance of them for comparison both in the Pleistocene and the modern seas. It is also independent of any questions as to the limits of species and varieties, inasmuch as it depends on careful comparisons of the living and fossil specimens ; and by whatever names we may call these, their similarity or dissimilarity remains unaffected. We have at present no means of tracing this fauna, as a whole, farther back. Some of its members we know existed in the Pliocene and Miocene without specific difference ; but some day the middle tertiaries of Greenland may reveal to us the ancestors of these shells, if they lived so far back, and may throw further light on their origin. In the meantime we can affirm that the lapse of time since the Pliocene

---

* See above, the remarks on the species of **Mya**.

has not sufficed even to produce new races; and the inevitable conclusion is that any possible derivation of one species from another is pushed back indefinitely, that the origin of specific types is quite distinct from varietal modification, and that the latter attains to a maximum in a comparatively short time, and then runs on unchanged, except in so far as geological vicissitudes may change the localities of certain varieties. This is precisely the same conclusion at which I have elsewhere arrived from a similar comparison of the fossil floras of the Devonian and Carboniferous periods in America.

A second leading point to which I would direct attention is the relative value of land ice and water-borne ice as causes of geological change in the Pleistocene. On this subject I have constantly maintained that moderate view which was that of Sir Roderick Murchison and Sir Charles Lyell, that the Pleistocene subsidence and refrigeration produced a state of our continents in which the lower levels, and at certain periods even the tops of the higher hills, were submerged, under water filled every season with heavy field-ice formed on the surface of the sea, as at present in Smith's Sound, and also with abundant ice-bergs derived from glaciers descending from unsubmerged mountain districts. These conclusions have been reinforced by the recent establishment of the fact of differential elevation and submergence, whereby the mountain ridges retained their elevation even when plains and table-lands were submerged. I need not reiterate the arguments for these conclusions, but may content myself with a reference to the changes of opinion on the subject. The glacier theory of Agassiz and others may be said to have grown till, like the imaginary glaciers themselves, it overspread the earth. All northern Europe and America

20

were covered with a *mer-de-glace*, moving to the southward
and outward to the sea. This great ice-mantle not only
removed stones and clay to immense distances, and
glaciated and striated the whole surface, but it cut out
lake basins and fiords, ground over the tops of the highest
hills, and accounted for everything otherwise difficult in
the superficial contour of the land. It was even transferred to Brazil, and employed to excavate the valley of
the Amazon. But this was its last feat, and it has
recently melted away under the warmth of discussion
until it is now but a shadow of its former self. I may
mention a few of the facts which have contributed to
this result. It has been found that the glacial boulder-
clays are in many cases marine. Cirques and other
alpine valleys, once supposed to be the work of glaciers,
are now known to have been produced by aqueous denudation. Great lakes, like those of America, supposed to be
inexplicable except by glacier erosion, have been found to
admit of being otherwise accounted for. The transport
of boulders and direction of striation have been found to
conflict with the theory of continental glaciation, or to
require too extravagant suppositions to account for them
in that way. Greenland, at one time supposed to be an
analogue of the imaginary ice-clad continent, has proved
to be an exceptional case which could not apply to the
interior of a wide continental area. The relation of
Greenland to Baffin's Bay and Davis straits is indeed
similar to that which may have obtained between the
Laurentide hills and the submerged valley of the St.
Lawrence, or to that of the Cordillera range to seas lying
west and east of it. The conditions of modern Greenland,
in short, at that time spread southward over the high
ridges exposed to the vapour-laden atmosphere of the

submerged continental areas, and the greatest of these analogues of Greenland was, no doubt, the Cordilleran system of glaciers depicted in the map prefixed to this chapter.

It has been the practice of the more extreme glacialists to attribute to their opponents the idea that all glacial work is done by icebergs, whereas they should have known that seas loaded with icebergs imply land covered with snow and ice. Iceberg-work implies glacier-work. It is these glacialists who have persisted in confounding the work of land-ice, icebergs and field-ice, in mixing up the earlier and later drifts, in neglecting the effects of the great movements of elevation and depression which were going on throughout the Pleistocene period, in omitting to consider the effects of the comparatively rapid movements of this kind which must have taken place from the crust suddenly giving way under tension, in confounding deposits obviously, from their structure and fossils, marine, with glacier moraines, in quietly assuming for glaciers an extension physically impossible, in neglecting to consider the possibility of tracts of verdure inhabited by animals on the margin of snow-clad hills and table-lands, in exaggerating the eroding and transporting power of glaciers, and minimizing that of sea-borne ice, and generally in misunderstanding or misrepresenting the glacial work now going on in the arctic and boreal regions. These are grave accusations, but I find none of the memoirs or other writings of the current school of glacialists free from such errors; and I think it is time that reasonable men should discountenance these misrepresentations, and adopt more moderate and rational views.

The facts indicate that there was an earlier and later period of glacial action and dispersion of boulders, that between these, in the middle Pleistocene period, large portions of the northern parts of the Northern Hemisphere possessed a climate not much colder than that enjoyed at present, and that in the height of the cold period only a limited portion of the north-east of Europe, the Alpine regions, the Cordillera of North America, the Laurentide hills and the Appalachians were deeply ice-capped, while the ice was flowing on all sides, north as well as south, into submerged areas.

In so far as Canada is concerned—and Canada includes the northern half of the American continent, the greatest of all the theatres of glacial action—the history of the Pleistocene period, as stated in the previous chapters, may be summed up as follows, beginning with the continental period of the newer Pliocene :—

1. In Canada and the eastern part of North America generally, it is universally admitted that the later Pliocene period was one of continental elevation, and probably of temperate climate. It is also evident, from the raised beaches holding marine shells, extending to elevations of 600 feet, and from boulder-drift reaching to a far greater height, that extensive submergence occurred in the middle and later Pleistocene. This was the age of the marine Leda clays and Saxicava sands found at heights of 600 feet above the sea in the St. Lawrence valley nearly as far west as Lake Ontario. It was also the time of the extensive drift over the great area of the western plains.

2. It is reasonable to conclude that the till, or boulder-clay, under the Leda clay, and its equivalents, belongs to the intervening period of probably gradual subsidence of the lower lands, accompanied with a severe climate and with snow and glaciers on all the higher grounds, sending glaciated stones into the sea. This deduction agrees with

the marine shells, bryozoa, and cirripedes found in the boulder-deposits on the lower St. Lawrence, with the unoxidized character of the mass, which proves subaquatic deposition, with the fact that it contains soft boulders, which would have crumbled if exposed to the air, with its limitation to the lower levels and absence on the hillsides, and with the prevalent direction of striation and boulder-drift from the north-east.

3. All these indications coincide with the conditions of the modern boulder-drift on the lower St. Lawrence and in the arctic regions, where the great belts and ridges of boulders accumulated by the coast-ice would, if the coast were sinking, climb upward and be filled in with mud, forming a continuous sheet of boulder-deposit similar to that which has accumulated and is accumulating on the shores of Smith's sound and elsewhere in the arctic, and which, like the older boulder-clay, is known to contain both marine shells and drift-wood.*

4. The conditions of the deposit of till diminished in intensity as the subsidence continued. The gathering ground of local glaciers was lessened, the ice was no longer limited to narrow sounds, but had a wider scope as well as a freer drift to the southward, and the climate seems to have been improved. The clays deposited had few boulders and many marine shells; and to the west and north there were deposits of land plants, and on land elevated above the water peaty deposits accumulated.

5. The shells of the Leda clay indicate depths of less than 100 fathoms. The numerous foraminifera, so far as have been observed, belong to this range, and I have never

* For references, see Royal Society's Arctic Manual, London, 1875. Fielden, Paper on Grinnel Land. Proc. Royal Socy. Dublin, 1878.

seen in the Leda clay the assemblage of foraminiferal forms now dredged from 200 to 300 fathoms in the Gulf of St. Lawrence.

6. I infer that the subsidence of the Leda clay period and of the interglacial beds of Ontario belongs to the time of the sea beaches from 450 to 600 feet in height, which are so marked and extensive as to indicate a period of repose. In this period there were marine conditions in the lower and middle St. Lawrence and in the Ottawa valley, and swamps and lakes on the upper Ottawa and the western end of Lake Ontario; and it was at this time that the plants described in previous pages occupied the country. It is quite probable, nay, certain, that during this interglacial period re-elevation had set in, since the upper Leda clay and the Saxicava sand indicate shallowing water, and during this re-elevation the plant-covered surface would extend to lower levels.

7. This, however, must have been followed by a second subsidence, since the water-worn gravels and loose, far-travelled boulders of the later drift rose to heights never reached by the till or the Leda clay, and attained to the tops of the highest hills of the St. Lawrence valley, 1,200 feet in height, and elsewhere to still greater elevations. This second boulder drift must have been wholly marine, and probably not of long duration. It shows little evidence of colder climate than that now prevalent, nor of extensive glaciers on the mountains; and it was followed by a paroxysmal elevation in successive stages till the land attained even more than its present height, as subsidence is known to have been proceeding in modern times.

8. For the region between the great lakes and the Rocky mountains and for the Pacific coast the sequence

is similar, but there was a greater amount of differential elevation as between the mountains and the plains. In the mountainous regions of the west, also, more especially in the interior of British Columbia, the evidence of great local glaciers is much more pronounced than on our lower mountains of the east.

I shall not attempt to extend these generalizations to the country south of the Canadian border, but must respectfully warn those of my geological friends who insist on portentous accumulations of land-ice in that quarter, that the material cannot be supplied to them from Canada. They must establish gathering-grounds within their own territory.

Note on Recent Papers.

While this work was in the press the discussion of questions relating to the glacial period in the United States and Europe has been actively proceeding. Sir Henry Howorth has treated the subject in an almost exhaustive manner in his work the "Glacial Nightmare," in which his point of view is very nearly that of the present work; though not like this confined to the case of Canada. Many important memoirs have also appeared in American and British periodicals, and in those of the Continent of Europe. Of these I shall notice only the following, as bearing closely on the scope of the previous pages :

Prof. Bonney, F.R.S., in a paper read before the Royal Geographical Society,* discusses in detail the question of glacial erosion, and arrives at the same conclusion which I stated in 1866, after visiting the Savoy Alps, viz., that

* "Nature," March 30, 1893.

glaciers are "agents of abrasion rather than erosion," and
that in the latter their power is much inferior to that of
fluviatile action.   Nor are glaciers agents in the excavation of lake basins, which are to be accounted for in other
ways; and the great gorges and fiords which have been
ascribed to them are due to aqueous erosion when the
continents were at a high level, before the glacial age.

An interesting and thoughtful paper, by Warren
Upham, has appeared,* in which he institutes a comparison between "Pleistocene and Present Ice-sheets."   The
present ice-sheets are stated to be four.   (1.) The Antarctic or that which fringes the Antartic continent and
is probably better entitled to the name than any other;
but which differs from the supposed ice-sheets of the
Pleistocene in fronting on the sea and discharging all its
produce as floating ice.   In this it certainly resembles
many of the great local glaciers of the Pleistocene.
(2.) The great nevé of Greenland, which, however, discharges by local glaciers, and these open on the sea, and
which has margins of verdure on its borders in summer.
(3.) The Malaspina glacier of Alaska, evidently a local
glacier of no great magnitude, though presenting some
exceptional features and showing the possibility of the close
contact of glacial phenomena and flourishing woodland.
(4.) The Muir glacier of Alaska, also a local glacier, but
perhaps, like the Malaspina, showing some features illus-
trative of local Pleistocene glaciers, more especially in its
apparent want of erosive power.

In the "conferences and comparisons," however, the
facts detailed in the earlier part of the paper are placed
in comparison with postulates respecting the Pleistocene

---

* Bulletin Geol. Society of America, March 24, 1893.

which are incapable of proof. (1.) It is taken for granted that the upper limits of glaciation in the mountain ranges of America indicate the thickness of a continental ice-sheet. They probably indicate only the upper limit of the abrasion of local glaciers. (2.) Hence it is computed that the thickness of a continental glacier flowing radially outward in all directions from the Laurentian highlands of Canada, amounted to two miles; and in connection with this it is stated that the maximum thickness of the great Cordilleran glacier of British Columbia has been estimated to have been about 7,000 feet; an entirely different thing, and referring to the maximum depth of a local glacier traversing deep valleys. (3.) It is admitted that the assumed continental glacier could not move without an elevation of the Laurentian highlands to the height of several thousand feet, of which we have no evidence, for the cutting of the deep fiords referred to in this connection must have taken place in the time of Pliocene elevation of the continents before the glacial period. (4.) The Upper and Lower Boulder drift, so different in their characters, are accounted for on the supposition that the former comes from material suspended in the ice at some height above its base, the other from that in the bottom of the ice. In like manner the widely distributed interglacial beds holding remains of land plants of North temperate character, are attributed to such small local occurrences of trees on or under moraines as appear in the Alaska glaciers. (5.) The rapid disappearance of the ice is connected with a supposed subsidence of the land under its weight, though from other considerations we know that if this was dependent on such a cause, it must have been going on from the first gathering of the ice, so that the required high land could not have existed. All the evidence, however, points to subsidence and elevation owing to other and purely terrestrial causes, and producing not produced by the glaciers of the Pleistocene.

It may be added that Upham accepts the recency of the glacial period, and its causation by changes of ocean currents, which of course would imply that its date coincided in Europe and America, though not necessarily or probably in the Southern Hemisphere.

The very important series of papers by Prof. Prestwich which have appeared within the last three years, and in which that veteran and able student of the later geological periods states his conclusions respecting the glacial and Post-glacial deposits of the South of England, contain a mine of information bearing on the glacial period in America. The papers by Hicks, Hughes, Lapworth, Mellard Reade, Nicholson and others, respecting the high-level gravels with marine shells in England and Wales, have also elicited facts which tend to bring them into harmony with those of America. The time was when the boulder-clays and raised beaches of Eastern America were explained by earthquake waves and glacier thrusts; but their vast extent and obviously submarine characters have rendered such contentions untenable, and it may be confidently predicted that this will be their fate in Great Britain also.

INDEX.

Erratum.—Page 195, line 4, for "chapter III." read "chapter II."

In Memoriam.—While this work was passing through the press, intelligence arrived of the death of my esteemed friend, Dr. John Rae, F.R.S., one of the most intrepid and successful of arctic explorers, and a diligent and accurate observer of the phenomena of nature in those dreary yet interesting regions with which he was so familiar. Some observations on shore ice, kindly contributed by him, are given at page 105 of this work.

www.ingramcontent.com/pod-product-compliance
Lightning Source LLC
Chambersburg PA
CBHW021504210326
41599CB00012B/1130